SRSK2R08

A Strategic Risk Communications Process for Outreach and Dialogue on Biosolids Land Application

by:

Sara Eggers, Ph.D.
Decision Partners, LLC.

Sarah Thorne
Decision Partners, LLC.

Gordon Butte
Decision Partners, LLC.

Katherine Sousa
Decision Partners, LLC.

2011

WERF

The Water Environment Research Foundation, a not-for-profit organization, funds and manages water quality research for its subscribers through a diverse public-private partnership between municipal utilities, corporations, academia, industry, and the federal government. WERF subscribers include municipal and regional water and wastewater utilities, industrial corporations, environmental engineering firms, and others that share a commitment to cost-effective water quality solutions. WERF is dedicated to advancing science and technology addressing water quality issues as they impact water resources, the atmosphere, the lands, and quality of life.

For more information, contact:
Water Environment Research Foundation
635 Slaters Lane, Suite G-110
Alexandria, VA 22314-1177
Tel: (571) 384-2100
Fax: (703) 299-0742
www.werf.org
werf@werf.org

This report was co-published by the following organization.

IWA Publishing
Alliance House, 12 Caxton Street
London SW1H 0QS, United Kingdom
Tel: +44 (0) 20 7654 5500
Fax: +44 (0) 20 7654 5555
www.iwapublishing.com
publications@iwap.co.uk

Library of Congress Catalog Card Number: 2010942713
Printed in the United States of America
IWAP ISBN: 978-1-84339-258-3/1-84339-258-5

Decision Partners, LLC.

ACKNOWLEDGMENTS

The project team would like to thank the Research Challenge SRSK3R08 Biosolids Risk Assessment Team, the Science Advisory Research Panel, City of Tulsa Public Works Department, Virginia Biosolids Council, Mid-Atlantic Biosolids Association, and the New England Biosolids & Residuals Association for their contribution to this work.

Research Team

Principal Investigators:
Sara Eggers, Ph.D.
Sarah Thorne, M.A.
Decision Partners, LLC.

Project Team:
Gordon Butte
Katherine Sousa
Patrick Gurian, Ph.D.
Drexel University

WERF Issue Area Team

Brain Bolte
City of Tulsa

Kari Fitzmorris, Ph.D.
Georgia Southern University

Roberta King, M.S.
King County Wastewater Treatment Division

Mike Moore
HDR, Inc.

Rebecca Parkin, Ph.D.
George Washington University

Anthony Pilawski
New Jersey Department of Environmental Protection

Amy Pruden, Ph.D.
Virginia Polytechnic Institute and State University

Joan B. Rose, Ph.D.
Michigan State University

Rick Stevens
U.S. Environmental Protection Agency

Lisa Williams
Synagro Technologies, Inc.

Water Environment Research Foundation Staff

Director of Research:	Daniel M. Woltering, Ph.D.
Program Director:	Alan Hais, P.E., BCEE

ABSTRACT AND BENEFITS

Abstract:

The long-term viability and sustainability of biosolids land application is dependent on continuously earning stakeholder confidence, trust and support. This requires ongoing and effective engagement and communication with key stakeholders about the beneficial use of biosolids in their communities. The goal of this risk communications research was to develop the process, methods, tools, and materials to enable biosolids managers to conduct effective outreach and dialogue with key stakeholders in the communities where they operate, on both biosolids land application and its beneficial use, and on specific challenges such as potential health risks of pathogens in biosolids.

Through this research, the science-based *Strategic Risk Communications Process for Outreach and Dialogue on Biosolids Land Application* was developed, which biosolids professionals can adapt as they design and conduct stakeholder outreach and dialogue in their local communities. The process, originally developed for Health Canada and the Public Health Agency of Canada, was modified to fit the unique (and often unmet) communications needs of biosolids professionals. Drawing on the key learnings from this project, the researchers developed a *Primer for Biosolids Professionals: Conducting Effective Community Outreach and Dialogue on Biosolids Land Application* (referred to as the Primer in this report) that offers step-by-step guidance on how to adapt and implement the process. It provides supporting tools and sample materials that biosolids professionals can customize for their own use.

The process was applied and validated through case study collaborations with two groups of biosolids professionals. As part of the process, in-depth research was conducted with local landowners who receive biosolids, neighbors to biosolids land application sites and, in one case study, community health officials. The research showed that these stakeholders want to learn more about local biosolids use and want the opportunity to discuss their interests, priorities and questions with biosolids professionals. They are willing to support local land application, but they want their questions and concerns about odor, safety and fairness to be appropriately addressed. Critical to their support is their trust and confidence in the people who produce, use and regulate biosolids to ensure the safety of themselves, their families, their neighbors and their environment. Actionable outreach and dialogue plans and pre-tested communications materials were developed based on the in-depth research findings and the needs of the specific biosolids programs.

Benefits:

- Demonstrates the need for biosolids professionals to more proactively and effectively communicate with community members and engage them in dialogue about the beneficial use of biosolids in their communities.
- Provides biosolids professionals with valuable hands-on tools and guidance to help them better communicate with stakeholders in their communities.
- Demonstrates, through two case studies, how the tools and guidance were applied by biosolids professionals to support outreach and dialogue in their local communities.

- Provides supporting tools and sample materials that biosolids professionals can customize as they develop their own outreach and dialogue efforts.

Keywords: Biosolids, biosolids land application, stakeholder engagement, risk communications, outreach and dialogue, stakeholder judgment, pathogens.

TABLE OF CONTENTS

LIST OF TABLES

LIST OF FIGURES

EXECUTIVE SUMMARY

The long-term viability and sustainability of biosolids land application requires community stakeholders' confidence, trust, and support for biosolids professionals and their operations, products, and services. Stakeholder judgment of the appropriate use of biosolids – producing and land applying biosolids in ways that maximize the benefits, while ensuring minimal risk to people and the environment – is fundamental to the future of biosolids operations across North America and in other countries.

In 2008, the Water Environment Research Foundation (WERF) and its subscribers recognized the need to develop systematic tools and guidance – building on the state of the science in risk assessment and risk communications – to help biosolids professionals more effectively manage and communicate the benefits and risks associated with biosolids land application. The Research Challenge provided an opportunity to develop risk analysis and risk communications tools that can help biosolids professionals improve management of biosolids land application and stakeholder judgment of its acceptability as a practice. The goal was to improve decision making on the appropriate use of biosolids by all potential users at the local, state and national level.

To meet this need, the researchers adapted their state-of-the-science Strategic Risk Communications Process (Thorne et al., 2006), along with supporting tools and materials that can be customized and used by biosolids professionals to meet their needs. Strategic Risk Communications is defined as the purposeful process of skillful interaction supported by appropriate information. Its purpose is to enable well-informed decision making and action on risks. It includes all communications messages and interactions that can influence the judgments, decisions and actions of both decision makers and stakeholders. This process, based on the Strategic Risk Communications Framework and Handbook (Thorne et al., 2006) that the researchers developed for Health Canada, was specifically referenced and called for in the initial Request for Proposals issued by WERF in the fall of 2007.

This risk communications project was conducted as part of a broader initiative to develop a risk assessment tool for assessing and communicating the risks of pathogens in biosolids land application. The researchers worked in close collaboration with a team of researchers, referred to as the Biosolids Pathogen Risk Assessment Team led by Dr. Patrick Gurian of Drexel University. Their report, entitled *Site Specific Risk Assessment Tools for Land Applied Biosolids* (SRSK3R08) presents a linked environmental dispersion, exposure, and health effects model to assess the relative risk of human infection from direct or indirect exposure to land-applied biosolids.

Specific objectives for the risk communications project included:

1. Integrating expert knowledge on biosolids land application management, developing a model to clearly describe the biosolids management as a complex "decision-making system".
2. Through two case study applications of the Strategic Risk Communications Process, eliciting the values, priorities and informational needs of landowners and near neighbors (and in one case study, health officials) in the host communities.

3. Using the case study research results to develop prototypical risk communications tools and materials that can be adapted by utilities and other stakeholders who are responsible for managing biosolids risk assessment and risk communications. These include:
   a. Guidelines and a step-by-step process for adapting the Strategic Risk Communications Process to fit the needs of biosolids professionals.
   b. A template for developing risk communications plans and materials (documents, training materials and support mechanisms) for biosolids professionals to enable effective implementation of risk communications and measurement of the effectiveness of those plans.
   c. Targeted risk communications messages for community members, municipal planners and others to enable sound public judgment of the health, environmental, economic and social impacts of biosolids land use.
4. Using the research results to inform the technical pathogen risk analyses and tool development by the Biosolids Pathogen Risk Assessment Team.

The Strategic Risk Communications Process has been adapted to address the unique needs of biosolids professionals and the circumstances that define the communications challenges and opportunities in the communities in which they operate. The process was tested and adapted through two case studies conducted in collaboration with the City of Tulsa Public Works and the Virginia Biosolids Council. These case studies demonstrated that the Strategic Risk Communications Process can be readily applied by biosolids professionals as they develop customized plans and tools to support outreach and dialogue efforts in their communities. The key learnings, plans, and materials developed from the case studies provide concrete examples of how the steps are implemented.

This research should significantly improve WERF subscribers' ability to design and conduct effective risk communications, primarily through outreach and dialogue with their stakeholders – the people in the communities where there is biosolids land application – supported by information that is relevant to their stakeholders.

Overview of the Process

The Strategic Risk Communications Process offers a set of six integrated steps and procedures that guide the implementation of critical tasks (See Figure ES-1.)

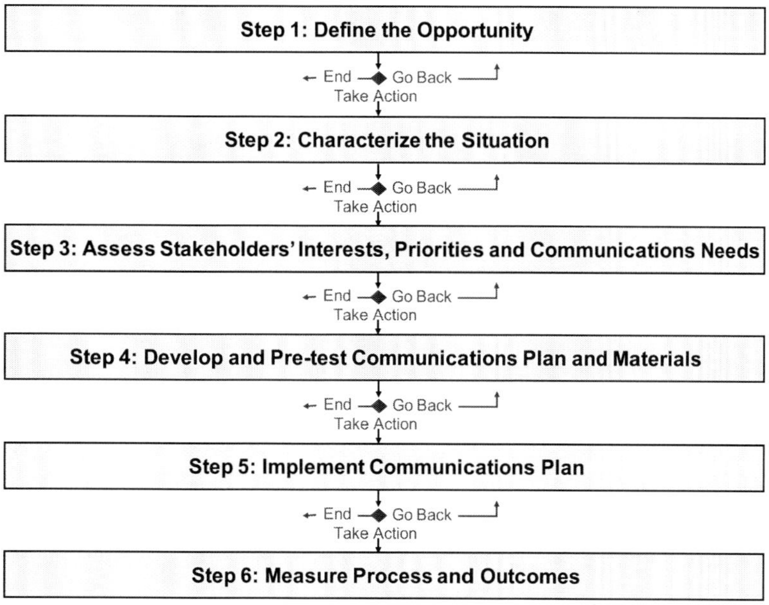

Figure ES-1. Strategic Risk Communications Process for Outreach and Dialogue on Biosolids Land Application.

- **Step 1: Define the Opportunity:** A Biosolids Communications Team, sponsored (and perhaps led) by the program manager, is formed. The team develops an Opportunity Statement – a clear statement of the current community outreach and dialogue opportunity for the biosolids program, including its goals, objectives and expected outcomes. The team also develops a plan for implementing the process, and (optional) guiding principles for community engagement and dialogue.

- **Step 2: Characterize the Situation:** The team conducts a preliminary technical assessment to understand the necessary scope and context of the biosolids land application program in the community(ies), defining the various related technical, social, regulatory, environmental, economic, and safety considerations. The team also conducts a stakeholder hypothesis to determine who the key stakeholders are, and the team's current understanding of those stakeholders' interests and priorities related to biosolids land application in their community.

- **Step 3: Assess Stakeholders' Interests, Priorities, and Communications Needs:** Using a variety of methods and tools – formal and informal research – the team gains an in-depth and first-hand understanding of key community stakeholders':

  - Values, priorities and interests.
  - Perceptions of biosolids and biosolids land application.
  - The factors that influence their judgment and decisions on the acceptability of biosolids land application in their community, and their resulting actions.
  - Communications needs: What they want to know and how they want to learn and engage in dialogue about biosolids land application?

  This understanding is compared to what was learned in Step 2 (Technical Assessment and Stakeholder Assessment) and the team defines the gaps and alignments.

- **Step 4: Develop and Pre-test Communications Plan and Materials:** Using the insight generated in the previous steps, the team develops a Communications Plan for community outreach and dialogue that addresses what the team learned about people's priorities, perceptions and communications needs.
- **Step 5: Implement Communications Plan:** The team implements the Plan developed in Step 4. They monitor, adjust and refine the plan and materials as required, based on what they learn through their outreach efforts about stakeholders' interests and priorities.
- **Step 6: Measure Process and Outcomes:** The team measures the effectiveness of their outreach and dialogue efforts, that is, the degree to which they achieve the Opportunity defined in Step 1. They also assess the effectiveness of the process and make recommendations for future team efforts as part of their commitment to continuous learning and continuous improvement.

The Strategic Risk Communications Process can provide value to biosolids professionals across a wide range of challenges, opportunities, situations and circumstances:

- It can be **tailored to fit the unique needs** of a community and/or a specific issue or challenge with key community stakeholders. These stakeholders may include landowners, people who live near land application sites, and other community officials.
- It can be **adapted** to address a specific project (e.g., a particularly sensitive land application site), or to provide an approach for an organization that conducts land application in many communities.
- It involves an **iterative process with integrated steps**.
- It can be **scaled to provide maximum value** with the available time and resources.
- It **supports continuous learning and continuous improvement.** Because it is a systematic process, it can be applied repeatedly, building knowledge upon knowledge, and tracking how the issues, priorities, perceptions and communications needs of your stakeholders change over time.

Applying the Process Through Case Studies

The Strategic Risk Communications Process was adapted to two case study applications. In the first case study (Appendices B and C), the researchers worked with a team from the City of Tulsa Public Works, called the Tulsa team. In the second (Appendix D), the researchers worked with a team comprising several members of the Virginia Biosolids Council, called the VA team. Key objectives of the case studies were to use the learnings from the research to adapt the process to the specific risk issues and challenges within these two host communities and to inform the development of prototypical strategies that can be adapted by utilities and other stakeholders.

The researchers worked with each case study team to identify the opportunities for outreach and dialogue, then conduct research to learn first-hand their community stakeholders' interests, priorities and communications needs regarding biosolids land application. Actionable plans and tested communications materials were developed based on the in-depth research findings and the needs of the specific biosolids programs. These materials included Guiding Principles for Community Outreach and Dialogue (see Appendix D), a dialogue presentation for use at community meetings (Appendix C-2), a brochure tailored to meet the needs of local community members (Appendix C-1), a set of more technical questions and references (Appendix C-3), and draft site signage that may be placed on site in advance of a land application (Appendix C-4). These materials can be useful for biosolids professionals to customize their own risk communication plans and materials to enable effective implementation of risk communications strategies and effective measuring of those strategies.

The case study teams saw real value in taking proactive steps to improve their current outreach and dialogue efforts. The process enabled them to: gain insight into their stakeholders' priorities and needs; understand at what level – who, what and when – to communicate; and learn how they could improve their engagement.

Research with Key Community Stakeholders

As a key component of the case study applications, research phone interviews were conducted and analyzed with: 12 landowners and 12 neighbors in Tulsa, Oklahoma; 24 neighbors in three Virginia communities (Fauquier County, Tidewater area and Lunenburg); and 6 Virginia Department of Health officials. The research followed an Expert Model/Mental Models research approach (Morgan et al., 2002). The method was designed by risk analysis and risk communications experts as a means to discover and characterize people's thinking and information needs about a complex issues in risk management and inform the development of focused communications plans and materials that address those priorities and needs. It plays an integral role in Health Canada's Strategic Risk Communications Process, and was used by the researchers to provide the foundational knowledge to adapt the process and apply it in the two case study demonstrations.

Chapter 3.0 provides details on the research design and preparation. Detailed research findings and a discussion on how they were incorporated into each case study communications plan are included in Appendix B, Tulsa Case Study Application of the Strategic Risk Communication Process, and Appendix D, Virginia Case Study Application of the Strategic Risk Communications Process. These research results can provide biosolids professionals some insight into the range of priorities and perceptions that may be shared by people in their own communities. They also provide in-depth insight that may provide additional context to prior research on public perceptions of biosolids land applications (e.g., Beecher et al., 2004; Krogmann et al., 2001; Robinson and Robinson, 2006.) A key difference between this research and much of the prior research is the focus is on specific populations, particularly near neighbors to land application sites, rather than a general population. (Note: These qualitative, small-sample research results should not be interpreted as being representative of the U.S. population. Use of percentages are used only to indicate the proportion of the sample that responded a particular way in the research.)

Most Tulsa landowners and neighbors demonstrated general awareness of biosolids and biosolids land applications. Interviewees in VA expressed a wider range of familiarity with biosolids. More broadly, interviewees, particularly neighbors in Tulsa and VA, generally did not appear to have an in-depth understanding of water and wastewater systems. Some commented that they were “not a part” of the wastewater system. When commenting on biosolids, interviewees, particularly those who appeared to be less familiar with biosolids, often drew on their perceptions of other environmental and health issues – and how these issues were managed by the people in charge. Examples included: the Gulf oil spill; PCB contamination in the Chesapeake Bay; fly ash and coal ash contamination on public land; animal manure; and commercial fertilizers.

Interviewees thoughtfully considered the perceived benefits, risks, costs and tradeoffs of biosolids land application in their communities. Near the end of the interview, interviewees were asked if they were *generally in favor* of biosolids land applications in their county, *generally not in favor* of biosolids land applications in their county, or *undecided.* All Tulsa landowners were in favor of biosolids land application. Tulsa neighbors were generally not in favor, citing the odor, health concerns, quality of life impacts, and uncertainty.

VA neighbors were more varied in their judgments. 50-60% of the VA neighbors in each community said they were "in favor", citing the value to the agricultural community; responsible management of a societal problem; and confidence in biosolids management. About 10% were not in favor, citing the unacceptable uncertainty about health and environmental risks. The remainder (about 30%) said they were undecided or in favor "with conditions". They wanted more assurance about the safety and long-term health and environmental impacts. VDH officials commented that they were in favor, generally referred to biosolids as the "best of both worlds".

The key factors that appear to influence (positively or negatively) interviewees' judgments of the acceptability of biosolids land application in their counties were:

- The interviewee's awareness and experience (or lack of) with biosolids.
- The interviewee's confidence (or lack of) in the safety of biosolids.
- The interviewee's experience (or lack of) with and perceptions of odor.
- The interviewee's concerns (or lack of) about contamination and/or concerns about the "unknowns".
- The interviewee's trust (or lack of) in the people who produce and apply biosolids, including the government officials who oversee and permit biosolids land application.
- The interviewee's perceptions and judgments of the agricultural and economic benefits to farmers and to the community.

Primer for Biosolids Professionals

A second key deliverable is a *Primer for Biosolids Professionals: Conducting Effective Community Outreach and Dialogue on Biosolids Land Application* (referred to as the Primer in this report). The purpose of the Primer is to give biosolids professionals (or biosolids communications teams) the tools they need to effectively implement the process.

The Primer was developed as stand-alone guidance document, available through WERF, to supplement this Research Challenge Report. It provides the principles of risk communications and guidance on the tasks and tools to use in each step of the process. It provides supporting worksheets that biosolids professionals can use to complete the step. Partially completed worksheets are included for each step in the Primer, draw on experience and insight gained from the WERF Research Challenge. (A set of blank worksheets is available in Appendix F of this report.) These worksheets and their supporting samples were developed with the intent that

Biosolids Communications Teams could adapt them as needed to fit their own needs, as part of working through the process. Finally, the Primer describes how the process can be scaled to provide maximum value with the available time and resources.

While the Primer, and its supporting tools, has been developed for biosolids professionals to use as they plan outreach and dialogue on biosolids land application, it can be applied to address the range of issues and opportunities that a biosolids program faces. Such opportunities may include developing strategies to communicate complex information on health or environmental topics, such as pathogens or emerging contaminants, or developing strategies to enable community support for new technologies and development, such as energy capture or composting facilities.

The Primer should to be of interest to all WERF subscribers, particularly utility managers and those who are responsible for communicating with stakeholders. It provides clear guidance to utilities that want to significantly improve their risk communications efforts. The contents of the Primer can provide the base for developing focused training for biosolids professionals on how to apply the process and develop and implement communications strategies and messages that address the specific interests and priorities of their stakeholders.

Applying the Process to Effectively Communicate About Pathogen Risk

Although the focus of this Primer was on developing broad stakeholder outreach and dialogue on biosolids land application, the process could also be used to address a number of challenges and opportunities facing biosolids professionals. This includes effectively communicating with stakeholders about pathogen safety. With the advances of risk assessment in pathogen risk of biosolids (i.e., the development of the microbial risk assessment tool by the WERF Research Challenge Biosolids Pathogen Risk Assessment Team (SRSK3R08)), there is a significant opportunity for WERF, its subscribers and partner organizations, to coordinate efforts in developing plans, guidance and supporting communications materials to facilitate effective dialogue on the pathogen safety of biosolids land application.

The process can be applied to help achieve this opportunity. One key component would be empirical research conducted with key stakeholders in order to understand, in depth, their priorities, interests and communications needs related to pathogen safety of biosolids land application. These stakeholders may include the farmers and near neighbors to an application site because they are the ones who are closest to any risks of pathogen exposure. Other key stakeholders might include utility and land applier employees and health department officials. The research design presented in this WERF Research Challenge could serve as a foundation. It could be tailored to include more specific questions about: what people think about pathogen risk in general, and specific to biosolids; how they compare pathogen risks in biosolids with other risks and what tradeoffs they are willing to take; what they consider to be "acceptable" risk of pathogens and why; what they expect from decision makers regarding the management of pathogen safety; their current confidence in pathogen safety and the factors that would increase their confidence in the safety of biosolids with regard to pathogens; and what additional information they want to have about pathogens and how they would like to receive that information.

Training and Dissemination

A key opportunity for WERF and its subscribers, along with members of other biosolids profession-related organizations, is to establish and conduct training for current and future biosolids professionals. The recommendation is to develop a comprehensive training package using the Primer as the training handbook. The purpose of the training would be to provide hands-on, in-depth training and coaching on how to operationalize the Strategic Risk Communications Process.

The Primer has been developed in a way so as to facilitate the development of training modules (e.g., through presentation slides). Participants would learn the essential principles, process and tools of Strategic Risk Communications, along with the rationale for their use. Hands-on training with coaching on the application of the methods and tools to a current challenge faced by each of the participants would be the focus of most of the training. Often one-on-one follow up coaching is recommended as the participants work through their case.

Training on the application the Strategic Risk Communications Process could follow with more focused dialogue skills training for biosolids professionals, and others on the team who will be conducting the outreach and dialogue. The dialogue training is typically a day-long session primarily focused on skills practice.

Key Observations from the Research Challenge

The Research Challenge demonstrated that long-term viability of biosolids land application depends on continuously earning stakeholder confidence, trust, and support. Over the course of this Research Challenge, it became clear that a fundamental opportunity for biosolids professionals – the people and organizations responsible for the safe production and appropriate use of biosolids – is to define and demonstrate a strong commitment to their people, their products and services, and the people in the communities where they operate. How? Through their management orientation. Management orientation is set of integrated beliefs and attitudes that guide the judgments, decision making and actions of members of the individual organizations that comprise what may be called the biosolids industry. Every day the biosolids professionals who comprise this industry demonstrate their management orientation through communications of all kinds – by what they say and what they do.

In conclusion to the Research Challenge, the researchers offers some key considerations on elements of a management orientation that can serve biosolids industry leaders and other biosolids professionals as they work to improve their communications, outreach and dialogue with their stakeholders. These considerations reflect the total experience of the researchers based on our interactions with a wide variety of biosolids professionals, as well as the specific research results of the WERF Challenge.

Specific recommendations for designing and conducting effective outreach and dialogue, based on the Strategic Risk Communications Process, can be found in the *Primer for Biosolids Professionals: Conducting Effective Outreach and Dialogue on Biosolids Land Application* (referred to as the Primer), which was a key deliverable of this Research Challenge and is available through the Water Environment Research Foundation.

While aligned with the practical recommendations in the Primer, the considerations below focus on the higher-value opportunities for biosolids professionals and their organizations. These include the following:

- Striving to systematically earn trust and, ultimately, societal support by actively demonstrating: a) your commitment to your people, products and communities; b) good performance; and c) your efforts to communicate effectively.
- Reframing biosolids as a product, not a waste. By reframing, the researchers mean redefining how you think about and communicate about biosolids – that is, biosolids are a valued product that provides a range of benefits to farmers, communities and society versus biosolids are a waste that society must deal with.
- Ensuring that good performance precedes your communications. As many industries and companies have learned the hard way, you can't communicate your way out of bad performance. Addressing performance issues – in operations, product quality, or the work of your people – must come before any communications about your performance as a company as demonstration of your efforts to continue to earn community support and your ongoing license to operate.
- Being proactive and leading the dialogue on biosolids land application.

With the development of a systematic process, actionable tools, step-by-step guidance, and in-depth demonstrations resulting from this Research Challenge, biosolids professionals should be much better prepared to design and conduct effective communication with their key stakeholders – the people whose decision making affects the success of their biosolids programs.

Helping biosolids professionals adopt the process as a leading practice is the next crucial step. To maximize the benefit of this approach, biosolids professionals will need to gain the necessary support, buy-in and resources from their key partners to fully implement the process and realize its benefits. For municipal utilities managers, as an example, this may require buy-in from municipal governments or state regulators.

Continued applications of the process will generate increasing value and position the process as a leading management practice for biosolids professionals. The process can also be applied and make significant contributions to water utility professionals' efforts on wastewater, stormwater and other environmental management programs.

Another opportunity to leverage this Research Challenge is to continue – through application of the process – to generate empirical insight into stakeholders' priorities, perceptions and communications needs on biosolids management. Using the case study applications as a base, similar research can be conducted in other communities, building a more complete picture of stakeholders (including the public's) current thinking on biosolids. The expert models/mental models research method is specifically designed to enable the integration of research across communities.

Finally, the work conducted as part of this Research Challenge, coupled with that of the development of the pathogen risk assessment tool by the Biosolids Pathogen Risk Assessment Team, can also support a continued effort by WERF, its partners and its subscribers, to advance a integrated risk management and risk communications approach to biosolids land application (Beecher et al., 2005). An integrated risk management and risk communications framework will become even more valuable as WERF and its partners continue to contribute to the development of other management tools and best practices, such as the Biosolids Odor Reduction Roadmap tool (BORR-03CTS9c) and incident response protocol (08HHE5PP).

CHAPTER 1.0

INTRODUCTION AND OBJECTIVES

1.1 Introduction

In the United States, approximately seven million dry tons of biosolids are produced annually by more than 16,000 municipal wastewater treatment facilities. Properly treated and managed, this wastewater by-product can be put to beneficial use through land application, as a soil amendment or soil replacement, rather than disposed of through landfills or incineration. Currently, approximately 55% of biosolids produced in the U.S. are land applied (North East Biosolids and Residuals Association, 2007).

The production, management and use of biosolids is a complex "system", supported and sustained by the decisions and actions of many different stakeholders. These include biosolids professionals, such as municipal utility staff, land appliers and other contractors who manage the production and use of a safe and effective product, as well as regulators who set standards and oversee their appropriate use. Importantly, it also includes stakeholders in the local communities where the biosolids professionals operate: farmers who benefit from this product; local officials who determine whether and how land application is acceptable in their community; and community members who give input on the acceptability of its use in their neighborhoods.

The long-term viability and sustainability of biosolids land application requires these stakeholders' confidence, trust and support for biosolids professionals and their operations, products and services. Stakeholder judgment of the appropriate use of biosolids is fundamental to the future of biosolids operations across North America and in other countries (NBP, 2005; Pepper et al., 2008). Simply put, biosolids professionals must earn, and continue to earn, their "license to operate".

Effective communication, which focuses on engaging and addressing the interests and priorities of stakeholders in the local communities, is essential to earning and sustaining that license to operate. The costs of ineffective communication can be very high, potentially resulting in the irreplaceable loss of the land applier's credibility; unnecessary, bitter, and prolonged public disputes on projects and costly delays; burdensome regulation; critical and unsupportive employees; and loss of commercial opportunities.

In 2008, the Water Environment Research Foundation (WERF) and its subscribers recognized the need to develop systematic and science-based tools and guidance to help biosolids professionals more effectively manage and communicate the benefits and risks associated with biosolids land application. As part of this Research Challenge, researchers adapted their state-of-the-science Strategic Risk Communications Process (Thorne et al., 2006). They developed and pre-tested supporting tools and materials that can be customized and used by biosolids professionals to meet their own needs.

Strategic Risk Communications is defined as "the purposeful process of skillful interaction supported by appropriate information" (Thorne et al., 2006). Its purpose is to enable well-informed decision making and action. It includes all communications messages and interactions – words and deeds – that can influence the judgments, decisions and actions of both decision makers and stakeholders.

The Strategic Risk Communications Process has been adapted to address the unique needs of biosolids professionals and the circumstances that define the communications challenges and opportunities in the communities in which they operate. The process was tested and adapted through two case studies conducted in partnership with the City of Tulsa Public Works and the Virginia Biosolids Council. These case studies demonstrated that the Strategic Risk Communications Process can be readily applied by biosolids professionals as they develop customized plans and tools to support outreach and dialogue efforts in their communities. The key learnings, plans and materials developed from the case studies provide concrete examples of how the steps are implemented.

This research should significantly improve the ability of biosolids professionals to effectively design and conduct communications with their key community stakeholders, and develop information that is relevant to their specific stakeholders.

1.1.1 Overview of Project Team

The authors of this report (referred to as the researchers) led the team that developed the Strategic Risk Communication Process adopted by Health Canada. They have adapted and applied the Strategic Risk Communications Process, and its supporting research methodologies, to a broad range of communications opportunities. They have contributed to the risk communications efforts of public health agencies, emergency responders, energy systems planners, chemical, manufacturing and mining industries, and water utilities, among others.

This risk communications project was conducted as part of an initiative to develop a risk assessment tool for assessing the risks of pathogens in biosolids land application. The researchers worked in close collaboration with a team of researchers, referred to as the Biosolids Pathogen Risk Assessment Team, led by Dr. Patrick Gurian of Drexel University, which included experts in biosolids, microbial risk, and risk assessment. The project also benefited from in-depth insight from the various members of the WERF Issue Area Team (IAT) charged with reviewing the Research Challenge project and the members of the case study teams. The researchers also drew on the insight and experience of a number of biosolids program managers, regulators, biosolids experts and others who comprised a Science and Regulatory Advisory Panel,[1] to support the Research Challenge. In this report, this set of collaborators and contributors to the development and application of an adapted Strategic Risk Communications Process for biosolids professionals is referred to as the project team.

1.1.2 Report Outline

This report describes how the Strategic Risk Communications Process was adapted as a tool for biosolids professionals and demonstrated through two case study applications. The remainder of this chapter provides an overview of the research methodology and project deliverables. Chapter 2.0 describes the background work (the first two steps of the Strategic Risk Communications Process, as applied to the Research Challenge) that supported the adaption of the process for biosolids programs. Chapter 3.0 describes the empirical research that was undertaken as part of the case study application of the process and summarizes the key learnings

[1] *See the accompanying Research Challenge SRSK3R08 Report:* *Site Specific Risk Assessment Tools For Land Applied Biosolids* *for the list of the Science and Regulatory Advisory Panel members.*

from that research. Chapter 4.0 presents the adapted Strategic Risk Communications Process for Outreach and Dialogue on Biosolids Land Application, discusses how the process was applied in the case studies, and discusses opportunities for training biosolids professionals. Chapter 5.0 discusses outreach efforts undertaken as part of the Research Challenge, and provides observations and future research directions.

1.1.3 Key Definitions

The following terms are used throughout this report:

- **Appropriate Use of Biosolids:** Stakeholder judgment that biosolids are produced and land applied in a way that maximizes the benefits, while ensuring minimal risk, of the land application to people and the environment.
- **Biosolids:** The primarily organic solid product yielded by municipal wastewater treatment that has been processed to meet the land-application standards in the Federal Part 503 Rule.
- **Biosolids Professionals:** The people working in municipal wastewater treatment facilities, land application companies and other organizations who are responsible for the management and use of land-applied biosolids.
- **Risk:** The chance of loss or injury. On a more technical level, risk is defined as a combination of the probability and severity of effects on organizational performance, health, property, the environment, or any other thing of value.
- **Risk Assessment:** The systematic process of estimating and evaluating risk. Risk estimation is the activity of measuring the likelihood (probability); type and severity of consequences in a risk situation; and any attendant uncertainty associated with the risk. Risk evaluation is the process by which risks are examined in terms of cost and benefits, and evaluated in terms of their acceptability, considering the needs, issues, and concerns of stakeholders.
- **Risk Management:** The systematic application of management policies, procedures and practices to the tasks of assessing, controlling and communicating risks.
- **Social Judgment:** People's thoughtful, considered judgments about issues that matter to them. It calls for a careful weighing of options and the costs, risks and benefits that characterize them.
- **Stakeholder:** Any individual, group or organization that may affect, be affected by, or perceive itself to be affected by decisions related to biosolids issues, projects, application, problems or policies.
- **Strategic Risk Communications:** The purposeful process of skillful interaction supported by appropriate information to enable well-informed decision making and action on risks. It is an integral part of the risk management process.

1.2 Research Objectives

The Research Challenge provided an opportunity to develop risk analysis and risk communications tools that can help biosolids professionals improve management of biosolids land application and stakeholder judgment of its acceptability as a practice. **The goal was to**

improve decision making on the appropriate use of biosolids by all potential users at the local, state, and national level.

1.2.1 Pathogen Risk Assessment Objective

A key objective of the Biosolids Pathogens Research Challenge was to develop and facilitate the use of risk assessment methodologies specific to microbial risk in biosolids. This objective was achieved through our Biosolids Pathogen Risk Assessment collaborators' work on the Research Challenge. Their report, entitled *Site Specific Risk Assessment Tools For Land Applied Biosolids* presents the Spreadsheet Microbial Assessment of Risk: Tool for Biosolids ("SMART Biosolids"). SMART Biosolids is a linked environmental dispersion, exposure, and health effects model to assess the relative risk of human infection from direct or indirect exposure to land-applied biosolids. SMART Biosolids assesses risk associated with land-applied biosolids through various exposure pathways (e.g., through consumption of surface water contaminated by runoff from a land application site). The SMART Biosolids model enables wastewater utilities, land applicators and regulators to estimate the exposure to pathogenic microorganisms – and resulting health risks – from biosolids land application under a variety site-specific conditions, such as setback distances, soil types, application rates, etc. The results of such analyses can yield insights into effective management and communications strategies.

The development of SMART Biosolids was conducted in close collaboration with the risk communications researchers, as part of an integrated risk assessment and risk management process.

1.2.2 Risk Communications Objective: A Strategic Risk Communications Process for Outreach and Dialogue on Biosolids

The primary risk communications objective was to support WERF subscribers in adopting the Strategic Risk Communications Process by customizing it to the specific (and often unmet) needs of WERF subscribers' biosolids land application programs. The process, originally developed for Health Canada and the Public Health Agency of Canada (Thorne et al., 2006) was specifically referenced and called for in the initial Request for Proposals issued by WERF in the fall of 2007. **The goal was to support biosolid professionals to significantly improve communications with stakeholders on the benefits and risks of biosolids land application, and in so doing, enabling stakeholders' judgments on the beneficial use of biosolids in their communities.**

This objective was accomplished through the adoption and application of the process to two case studies. The approach was refined for biosolids land application programs and guidance, and prototypical strategies and materials were developed to help biosolids professionals adapt the process to the challenges they face in their own biosolids programs.

In conducting the initials steps of the process ("Define the Oppportunity", and "Characterize the Situation", described in more detail in Chapter 2.0) in summer and fall 2008, it became clear that the risk communications issues and challenges facing biosolids professionals are complex and extend beyond the issues of biosolids pathogen safety. The biosolids professionals and other experts consulted at this early stage of the Research Challenge stressed the necessity of providing a tool to understand and address what they believed were stakeholders' more pressing questions and concerns about odor, long-term impacts to humans, animals and the environment, and transparency in decision making. This was confirmed when the process was initiated with the two case study teams.

Consequently, the risk communications objectives of this Research Challenge were refocused to enable understanding of the full context on stakeholders' values, interests, priorities and information needs related to biosolids land application. As research and experience shows, without this fuller context, more focused communications on a specific topics, such as pathogens, would be inadequate and/or risk missing the mark. Biosolids professionals could find themselves communicating with people about topics that they are not currently thinking about and/or providing information that is not relevant to people and not useful to them as they make decisions about the acceptablity of biosolids land application. A key principle of effective risk communications is to build on people's information needs, which calls for conducting empirical research to discover what people know, don't know, misunderstand, want to know, and who and what communications processes they trust.

Specific objectives for the risk communications project included:

1. Integrating expert knowledge on biosolids land application management, developing a model to clearly describe the biosolids management as a complex "decision-making system".
2. Through two case study applications of the Strategic Risk Communications Process, eliciting the values, priorities and informational needs of landowners and near neighbors (and in one case, health officials) in the host communities.
3. Using the case study research results to develop prototypical risk communications tools and materials that can be adapted by utilities and other stakeholders who are responsible for managing biosolids risk assessment and risk communications. These include:
   a. Guidelines and a step-by-step process for adapting the Strategic Risk Communications Process to fit the needs of biosolids professionals.
   b. A template for developing risk communications plans and materials (documents, training materials and support mechanisms) for biosolids professionals to enable effective implementation of risk communications and measurement of the effectiveness of those plans.
   c. Targeted risk communications messages for community members, municipal planners and others to enable sound public judgment of the health, environmental, economic and social impacts of biosolids land use.
4. Using the research results to inform the technical pathogen risk analyses and tool development by the Biosolids Pathogen Risk Assessment Team.

1.3 The Opportunity to Conduct More Effective Outreach and Dialogue

1.3.1 Social Judgment on Appropriate Biosolids Land Application

Biosolids risk management researchers and planners, including WERF, have identified significant challenges associated with individual, group or community objections to biosolids land application (Beecher et al., 2005; Pepper et al., 2008). These challenges, which can be characterized as "social friction" (Carpenter, 2009; Thorne and Butte, 2007), are not unique to biosolids programs. They are felt by planners and managers who support a wide range of social and environmental projects and technologies, even when their proponents believe that the project or technology yields significant societal benefit. Examples include wind energy development

(Wolsink, 2000; Devine-Wright, 2005), electricity transmission (Vajjhala and Fischbeck, 2006), hazardous waste facilities (Kuhn and Ballard, 1998), mental health facilities (Takahashi and Dear, 1997), penal facilities (Shichor, 1992), and suburban redevelopments (Simonson, 1992).

The social friction that arises in project planning or operations is due to complex differences in perceptions, values and understanding of interdependent stakeholders who believe they have a stake in the decisions being made. Note: In this report, the term social friction is used in lieu of more common terms such as "Not-In-My-Back-Yard" (NIMBY). The researchers argue that such terms are not helpful and may even jeopardize constructive dialogue in planning or management (Wolsink, 2006). They focus one-sidedly, and most often pejoratively, on individuals or groups rather than the underlying issues (values, risks, benefits) that are at the crux of the issue at hand. In doing so, they may trivialize the interests and priorities of the individuals affected by the issue. These terms also ignore the relationships between multiple stakeholders in the system, including community members, and their potential role in the successful resolution of the issue and opportunities related to the proposed project.

Social friction, properly channeled, can play an important role in ensuring due consideration to the value, benefits and risks that projects may bear on society, and in particular the people who may be most affected by that project. Thus, successful introduction and application of a technology, such as biosolids for land application, requires *social judgment* of the acceptability of the benefits, risks and costs of that technology. Social judgment refers to people's thoughtful, considered judgments about issues that matter to them. It requires shared understanding of the biosolids professionals and their stakeholders of the project need, benefits, costs and risks, and how stakeholders value outcomes and make tradeoffs.

Social judgment is distinct from public opinion. Public opinion can be thought of as people's top of mind, "no-sacrifice-for-me" views, often developed on the basis of hypothetical situations, with limited context, incomplete information and without consideration of a complex set of factors and tradeoffs. Social judgment, on the other hand, tends to emphasize coherence and integration of views. It represents a weighing of pros and cons inherent in options and a reconciliation of them.

Social judgment has been shown to result from many factors: people's understanding of the need or opportunity, their trust in the professionals, the process, and the technology, and, often most importantly, the degree to which they have, and believe they have, meaningful input into the decision-making process (Gregory et al., 2003).

1.3.2 Understanding and Addressing Stakeholders' Communications Needs

Communications that can enable and focus social judgment are the key to minimizing and managing social friction. Thus, broad social engagement of all stakeholders is a critical component of a successful planning project and requires shared understanding of the project need, benefits, costs and risks. It is the proactive involvement – through dialogue – of community stakeholders in a way that is meaningful and relevant to them, and is respectful of their information needs, objectives and values regarding the process. It affords biosolids professionals the opportunity to listen to and learn from their stakeholders. And it enables stakeholders to thoughtfully weigh the benefits, risks and tradeoffs of biosolids operations, including land application, in their communities and decide on their acceptability.

However, research and experience have shown that experts' beliefs about what to communicate on issues, to whom, and how, can represent barriers to effective communication.

For example, experts often predict that laypeople just need to be "educated" or worse, that they are "irrational" about risks and, therefore, will be inherently difficult to "reason" with (Thorne et al., 2006). Perhaps because of this, biosolids professionals have historically used a "decide-and-defend" approach to communications. As described by Beecher et al. (2004), "the biosolids manager, in consultation with the biosolids user, regulatory staff, and agricultural advisors, makes the decisions... Once the decisions are made, the public just needs to be educated, so they will accept the decisions (page 46)."

However, over the last 30 years, it has become evident that speculation about people's interests and priorities is naïve at best and often risky. Lay people typically address decisions from a different conceptual framework than experts, and use different terms. They often want and expect to participate in the process of defining options and making tradeoffs. When asked to think systematically about a problem or opportunity, they typically demonstrate in-depth thinking, thoughtfulness, richness and subtlety and great "common sense". Most people are in touch with the need for relevant and credible information, the consequences of various decisions, and how options may play out with respect to their interests and priorities and those of the broader community.

1.3.3 Building on Prior Research on Public Perception of Biosolids

A WERF Research Challenge 2004 Report entitled "Public Perception of Biosolids Recycling: Developing Public Participation and Earning Trust" provides a comprehensive review of the current literature on public perception of biosolids. Reflecting on their key learnings from their review (described in Chapter 2.0), Beecher et al. (2004) stated:

> "Over the course of this project, the investigators changed their perspective from a focus on finding ways for biosolids managers to gain 'public acceptance' of their biosolids recycling programs to a focus on 'developing public participation and earning trust'." (p. I-15).

Beecher et al. offered a number of recommendations that confirmed the need for a strategic approach to dialogue and outreach with community stakeholders:

- ♦ Success of land application projects is dependent on stakeholder outreach and engagement, and communications.
- ♦ The goal of communications is to build public trust, credibility, and confidence in the biosolids land application program.
- ♦ Communicating early is important for all stakeholders but especially for those who perceive they will be most affected by the project (i.e., near neighbors, farmers, local community).
- ♦ A stepwise-approach to engaging stakeholders is best in order to gain trust and build support.
- ♦ Engagement and dialogue should be undertaken by experts and facilitators who have the needed scientific and communications skills.

These key learnings were fundamental to setting the stage for the adaptation of the Strategic Risk Communications Process to meet these needs.

1.4 Overview of Research Challenge Approach and Research Method

1.4.1 Strategic Risk Communications Approach

With the 2004 Report as a base, WERF wanted to provide a science-based risk communications approach to biosolids land application challenges and opportunities. Ideally, the approach would be one that was proven and readily adaptable to the unique challenges of WERF subscribers.

The project team adopted a state-of-the-science integrated risk management process to effectively manage biosolids land use and, as an integral component of that process, the Strategic Risk Communications Process, as outlined in the Health Canada and the Public Health Agency of Canada's Strategic Risk Communication Handbook (Thorne et al., 2006). The process is comprehensive and is an integral part of the Health Canada and PHAC decision-making framework (Figure 1-1).

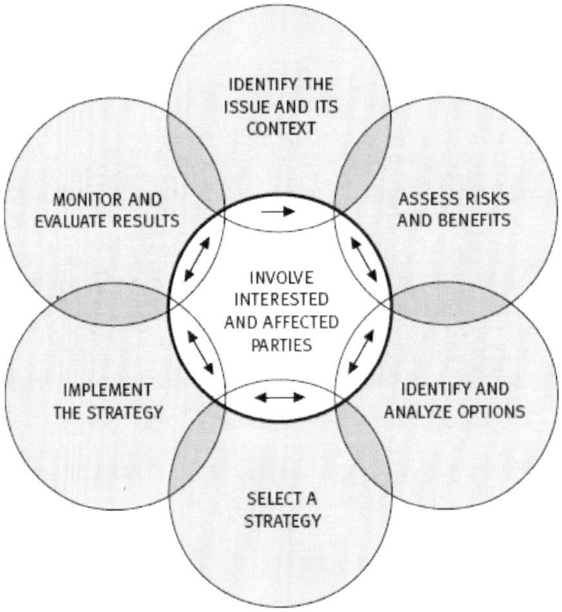

Figure 1-1. Health Canada's Decision-Making Framework (Health Canada, 2006, p.2-9).

The Strategic Risk Communications Process was designed as an integral part of current risk management processes, principally, the Canadian Standard Association's Q850 Risk Management: Guideline for Decision-Makers (CAN/CSA, 1997). This process is also aligned with the U.S. Presidential/Congressional Commission on Risk Assessment and Risk Management Process (1997) and was reviewed in the development of the International Organization for Standardization's (ISO) 31000 Guideline on Risk Management (ISO, 2009).

The Strategic Risk Communications Process, in Figure 1-2 provides step-by-step detail of the risk communications activities undertaken to support every step in the risk management process.

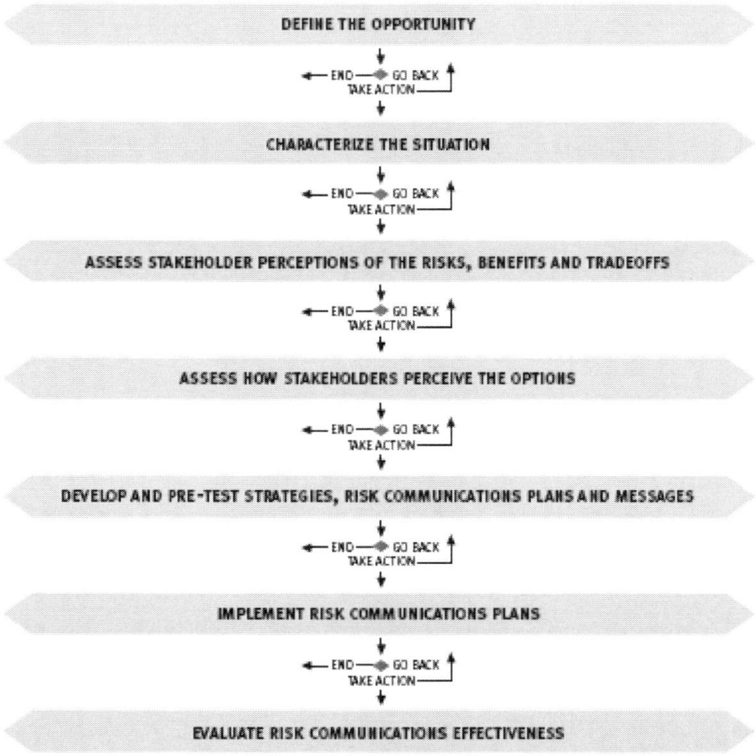

Figure 1-2. Health Canada's Strategic Risk Communication Process (Thorne et al., 2006).

Health Canada's Strategic Risk Communications Process has seven steps: 1) definition of the opportunity; 2) characterization of the situation; 3) assessment of stakeholder perceptions of the risks, benefits and tradeoffs; 4) assessment of how stakeholders perceive the options; 5) development and pre-testing of strategies, risk communications plans and messages; 6) implementation of risk communications plans; and 7) evaluation of risk communications effectiveness.

The Strategic Risk Communications Process taps current understanding in the relevant disciplines – risk management, risk perception, behavioral decision research, psychology and engineering. It was designed as an integral part of an effective risk management process.

Over the past two decades, the process, and earlier versions of it, have been adapted and applied to address a broad range of risk management and risk communications challenges and opportunities in public health, such as food and drug safety and occupational health and safety. It has also been adapted and applied to address the specific needs and opportunities facing a number of sectors, such as chemical, forest products, mining, power generation and transmission, and water utilities. Specific examples include:

- Helping the Canadian Council of Ministers of the Environment (CCME) Biosolids Task Group develop a Canada-wide approach for biosolids management by designing and developing key consultation documents for provincial and territorial biosolids consultation teams to use in their stakeholder consultation initiatives. The resulting Tool Kit, Resource Guide and Dialogue Presentation, were based on the Strategic Risk Communication Process developed for Health Canada (Thorne et al., 2006).

- Supporting the American Water Works Association Research Foundation in creating a systematic approach to better identifying, managing and communicating about emerging contaminants in drinking water and the risks associated with them (Parkin et al., 2004).
- Working with the Association of Power Producers of Ontario (APPrO), together with EPCOR Utilities Inc, a large Canadian water and energy company, and the Ontario Power Authority (OPA) to develop a sustainable approach to addressing the challenges of individual, group or community objections (i.e., social friction) to siting energy and/or industrial facilities or infrastructure (Thorne and Butte, 2007). The results of this work supported the development of methods and tools to help Canadian electricity planners and proponents effectively and proactively minimize the potential for social friction.

1.4.2 Expert Model/Mental Models Research Methodology

Effective communications strategies, including outreach and dialogue on biosolids, must reflect the best available (expert) understanding of the complex topic or management issue at hand. Equally important, they must be based on an in-depth understanding of stakeholder interests, priorities and information needs related to those topics. Thus, successful adoption of the Strategic Risk Communications Process should include sound and purposeful research with both experts and stakeholders. The expert model/mental models research method (Morgan et al., 2002) is ideally suited to generate needed insight in both the expert and stakeholder domains The method was designed by risk analysis and risk communications experts as a means to discover and characterize people's thinking and information needs about a complex issues in risk management and inform the development of focused communications plans and materials that address those priorities and needs. It plays an integral role in Health Canada's Strategic Risk Communications Process, and was used by the researchers to provide the foundational knowledge to adapt the process and apply it in the two case study demonstrations.

Over the past 30 years, researchers have used the mental models approach to address challenges presented by people's understanding of complex issues and processes, such as risk and risk management. The concept of mental models is a well-established theory in psychology and has been the focus of extensive research (Johnson-Laird, 1983; Bostrom et al., 1992; Atman et al.,1994; Fischhoff et al., 1997). A person's "mental model" can be thought of as a complex web of deeply – and often subconsciously – held beliefs, which affect how an individual defines an opportunity, reacts to information, forms judgments and makes decisions. These beliefs may be complete and correct, or they may have gaps that are consequential to decision making and action. Mental models tend to define the boundaries of thought and action and prevent people from seeing alternative perspectives. As such, they limit people to familiar patterns of reasoning and action.

The mental models research method seeks to understand people's mental models in order to directly inform the development of communications to best address those mental models. The method is well-established in the fields of risk analysis, decision research and risk communications (Downs et al., 2010). Through mental models research, one can discover:

- What people know that is correct and essential to making an informed decision.
- What they misunderstand that is consequential.

- What they do not know that is consequential.
- What they want to know that is important to them.
- What criteria they use to judge the trustworthiness and competence of people, organizations, and their communications processes and media.

The method entails in-depth research interviews designed and analyzed against a structured formulation of the complex topic at hand, in this case biosolids land application. The approach begins by integrating the knowledge of experts on a given topic. One method for integrating knowledge is through an *expert model*, which provides a formal representation of how situations are understood, capturing the range of legitimate opinion and uncertainty. Expert models are essential management and communication development tools. Done well, they capture and integrate the broad range of critical knowledge of a topic across the system involved in risk management and risk communications. An expert model also provides the analytical framework from which researchers can design, conduct, analyze and report on stakeholders' *mental models*. More detail on expert modeling is found in Chapter 2.0.

The mental models research method follows with empirical research into stakeholder thinking on the issues at hand. This research component involves individual, semi-structured interviews, with a well-defined group – called a cohort – of stakeholders. The interview is designed to address key topics identified in the expert model and allow for other topics to emerge through free expression. Structured qualitative analysis of the interviews against the expert models enables identification of the key areas of alignment and critical gaps in the thinking of experts and the stakeholders. It identifies the critical influences on people's decisions, including: their underlying priorities, how they conceptualize issues and opportunities, how they perceive the risks and benefits, and what tradeoffs they are willing to make. It also identifies what stakeholders want to know, and whom and what communications processes they trust. The insight generated by such research directly supports the development of targeted outreach and communications strategies. More detail on mental models research is found in Chapter 3.0.

The mental models research approach is among the most robust and adaptable of qualitative decision and communications research methods. It yields rich and high-quality data with small samples of individuals (Morgan et al., 2002). One of the key strengths of the mental models approach is its flexibility, allowing for remarkable integration with other research methods, including social marketing and community-based research approaches. Mental models research uncovers people's perceptions in depth, which can then provide critical insights to support development more structured research and evaluations. In addition, the expert model, which is used as the analytical framework for mental models research, can also be used to analyze the data from other forms of research. The mental models research approach also implicitly treats stakeholders, and their interests and priorities, with respect, thereby strengthening stakeholders' judgments of the trustworthiness and credibility of the sponsoring organization.

The mental models research approach has been used to guide risk communications strategies related to a wide range of topics. A few examples include: produce growers' adoption of good agricultural practices (Eggers et al., 2010); consumers readiness for threats of food terrorism (Eggers et al., in press); drinking water safety (Parkin et al., 2004); climate change (Palmgren et al., 2004), immunizations (Downs, Bruine de Bruin and Fischhoff, 2008), and many others.

1.5 Overview of Research Challenge Deliverables

The Research Challenge resulted in three key deliverables: an expert model, a Primer for biosolids professionals and two case study applications of the Strategic Risk Communications in two host communities.

1.5.1 Expert Model of Influences on an Integrated Risk Assessment and Risk Communications System for Appropriate Biosolids Land Application

A key deliverable of this project is the **Expert Model of Influences on an Integrated Risk Assessment and Risk Communications System for Appropriate Biosolids Land Application.** A narrative of the Expert Model is presented in Appendix A.

This Model integrates knowledge on key factors, such as regulations, environmental conditions and workforce quality that must be considered when developing biosolids land application risk assessment and risk communications tools. The Expert Model was developed in consultation with the project team, WERF, the IAT and the case study teams. The Expert Model Narrative provides an overview of the Expert Model and its development.

The expert model may be of interest to WERF subscribers, researchers and regulators as it can be readily adapted to fit the needs of individual land application programs. It may also be useful as a tool to communicate with key stakeholders (e.g., regulators, policymakers, etc.) on the numerous influences on successful biosolids land application.

1.5.2 Primer for Biosolids Professionals: Conducting Effective Outreach and Dialogue on Biosolids Land Application

A second key deliverable is a *Primer for Biosolids Professionals: Conducting Effective Community Outreach and Dialogue on Biosolids Land Application* (referred to as the Primer in this report). The purpose of the Primer is to give biosolids professionals (or biosolids communications teams) the tools they need to effectively implement the process.

The Primer was developed as stand-alone guidance document to supplement this Research Challenge Report. It provides the principles of risk communications and guidance on the tasks and tools to use in each step of the process, including:

- What the step is and why it is important.
- What the team will be able to do at the completion of the step.
- Tasks to be completed as part of the step.
- Tips on how to accomplish the tasks.
- Supporting worksheets that biosolids professionals can use to complete the step. Partially completed worksheets are included for each step. These samples draw on experience and insight gained from the Research Challenge. They were developed with the intent that Biosolids Communications Teams could adapt them as needed to fit their own needs, as part of working through the process. Blank worksheets are provided in Appendix F.
- How the process can be scaled to provide maximum value with the available time and resources.

While the Primer, and its supporting tools, has been developed for biosolids professionals to use as they plan outreach and dialogue on biosolids land application, it can be applied to address the range of issues and opportunities that biosolids programs face, including:

- Communicating effectively about real and perceived health risks associated with biosolids.
- Developing "Guiding Principles" and/or "Codes of Good Practices" for biosolids management.
- Developing strategies to communicate complex information on health or environmental topics, such as pathogens or emerging contaminants.
- Designing new programs for biosolids use, such as urban land-use projects.
- Enabling community support for new technologies and development, such as energy capture or composting facilities.

The Primer should to be of interest to all WERF subscribers, particularly utility managers and those who are responsible for communicating with stakeholders. It provides clear guidance to utilities that want to significantly improve their risk communications efforts. The contents of the Primer can provide the base for developing focused training for utility members on how to apply the process and develop and implement communications strategies and messages that address the specific interests and priorities of their stakeholders.

1.5.3 Case Study Applications of the Strategic Risk Communications Process and Prototypical Materials

The Strategic Risk Communications Process was adapted to two case study applications (summarized in Appendices B and D) in two host communities: Tulsa, OK and Virginia Biosolids Council. Key objectives of the case studies were to use the learnings from the research to adapt the process to the specific risk issues and challenges within these two host communities and to inform the development of prototypical strategies that can be adapted by utilities and other stakeholders.

As part of the case study applications, prototypical risk communications materials were developed. A sample of these materials developed from the Tulsa Case Study is presented in Appendix C. They include a pamphlet for local community members on Tulsa's biosolids land application, a neighbor dialogue presentation and speaking notes, more technical questions and references, and draft site signage that may be placed on site in advance of a land application. Biosolids professionals can customize these to suit their own risk communication plans.

CHAPTER 2.0

DEFINING THE OPPORTUNITY AND CHARACTERIZING THE SITUATION

2.1 Overview

The researchers recommended that the key tasks of the Research Challenge follow the Strategic Risk Communications Process. This served two purposes: a) to ensure that the resulting communications processes, tools and methods developed from the Research Challenge met the needs of biosolids professionals; and b) to demonstrate the value of the process as applied.

This chapter describes the tasks and key learnings from the first two steps of the Strategic Risk Communications Process: Defining the Opportunity and Characterizing the Situation.

2.2 Step 1: Define the Opportunity

As a first step, the researchers led the project team in a facilitated discussion to develop the Opportunity Statement, defining the project scope, process deliverables, and communications outcomes. At the first project team meeting, in August 2008, the project team developed the following Opportunity Statement:

The Opportunity of the WERF Biosolids Research Team is to build an integrated risk management/risk communications system comprising the development and pre-testing of state-of-the-science risk assessment and strategic risk communications tools in collaboration with key stakeholders. We will provide WERF subscribers with tools and training in their use starting in 2009.

Throughout the project, the Opportunity Statement was used to ensure alignment among the project team members, to guide the development of the Research Challenge workplan, and to communicate with key stakeholders about the Research Challenge.

2.3 Step 2: Characterize the Situation

For communications to be effective, they must start with shared understanding at the expert level of the context or system in which decisions about the acceptability of biosolids land application are made. Such understanding involves many complex issues, which must be systematically identified and addressed. Thus, Step 2 is a critical step of the process. It involves gathering and integrating the knowledge and insight on the complex topics of biosolids management. This includes characterizing in-depth the biosolids professionals' current understanding of their biosolids program, their stakeholders, and the community(ies) in which they operate.

The two primary activities of this step are *Technical Analysis* of biosolids management and a *Preliminary Stakeholder Analysis.* The purpose of the *Technical Analysis* is to integrate available knowledge on the biosolids risks and risk management system. One of the most effective ways to capture this understanding is by developing an expert model. As a formal integration of expert knowledge of a system, expert models are essential management tools. They enable identification of critical issues in risk assessment and risk management; gaps in current policies and procedures and opportunities to improve stakeholder engagement; and can be readily revised and refined over time to capture the most current understanding of key variables in the system it addresses. The development of expert models, typically through literature reviews and elicitation with experts from a range of disciplines, can be an effective way to build shared understanding of the system being addressed. An expert model also provides the analytical framework from which researchers can design, conduct, analyze and report on stakeholders' *mental models*. The researchers facilitated the integration of knowledge from a secondary literature review, from the expertise of the project peam and the IAT review panel as well as that from other experts and key stakeholders in biosolids management.

The second component of Step 2 is to conduct a *Stakeholder Hypothesis*. The goals of the Stakeholder Hypothesis were to: a) identify the range of stakeholders who had an interest in biosolids land application and biosolids risk management decisions; b) identify one to three key stakeholders to be the focus of the mental models research; and c) use insight from experts, documented case studies, scholarly literature and experience from similar projects, to gain empirical insight into the values, priorities, interests, perceptions, and information needs of these key stakeholders.

2.4 Literature Review

The researchers first reviewed scholarly articles, regulatory and guidance documents related to community and public perceptions of biosolids and biosolids risk management, and general research on risk communications. The review included an Internet search on topics related to biosolids land application and pathogen risk. The researchers consulted with individuals on the project team as needed.

2.4.1 Public Perception of Biosolids Land Application

In its 2004 WERF Biosolids and Residuals Research Challenge Report, Beecher et al. advanced the field's understanding about public attitudes toward biosolids and the factors that influence the ultimate success of biosolids projects. As groundwork, a comprehensive literature review gathered extensive documentation and breadth of opinion from peer-reviewed, published, and gray literature about two areas of interest: 1) public interactions with and attitudes about biosolids recycling and 2) significant developments in academic knowledge about public reactions, beliefs, and behaviors about similar environmental projects and issues. The literature review was paired with case studies of stakeholder interactions during fourteen biosolids projects. These reviewed a range of experience, from projects that created highly positive stakeholder relationships to projects that were stopped due to public protest.

Beecher et al. report on a 2002 study conducted by the University of New Hampshire Survey Center. The team conducted a nationwide survey of 1,069 randomly sampled homeowners and home renters (subjects thought to have some knowledge of agriculture and sewage issues). The survey showed little regional difference in perceptions or knowledge about biosolids. Generally, people had limited knowledge about biosolids. Only 14% had at least "a fairly good idea" about what biosolids are. Respondents were generally uneasy about the topic, and had questions about it. However, most believed that biosolids posed a relatively small risk to them.

Ninety-three percent responded positively to the idea of wastewater treatment. Respondents also responded favorably to other topics which touched upon environmental health and recycling, including "constructive uses of biosolids (creating energy, recycling nutrients)", "return[ing] nutrients to the soil", and "dispos[ing] of a necessary waste" (Beecher et al., 2004, p. *ES-2*). In contrast, people responded negatively to the term "sewage sludge", the concept of biosolids originating from a large city or industrial use, and especially their overall perception of uncertainty around biosolids. They said "the strongest argument against biosolids recycling is "not enough is known". (Beecher et al., 2004, p. ES-3).

When asked about a hypothetical situation where a neighbor uses biosolids, respondents said they trusted and would consult "friends and neighbors, government agencies, and academic researchers" as initial information sources. They preferred entities, such as government agencies, that were perceived as knowledgeable and objective. People distrusted sources that they believed had "a profit motive". And, "if they need to learn more about biosolids, people want more information about many different aspects" of the topic (Beecher et al., 2004, p. ES-2).

Research since the Beecher et al., report supports the earlier WERF project's conclusions. Robinson and Robinson (2006) found in a random-sample survey of 303 members of the general public in a large metropolitan area in Southeast U.S. that respondents generally lacked *a priori* awareness and knowledge on biosolids. However, when asked, respondents were generally favorable toward various biosolids use options (farmlands, grazing lands, public parks, highway medians, home gardens, lawns, forest lands, and use as mulch after composting), however the degree of support appeared to be related to the perceived risk of human exposure to biosolids. For example, forest land (4.24) and highway medians (4.23) received the highest mean attitude rating (out of a 5-point Likert scale) and home garden (3.18) received the lowest mean attitude rating.

2.4.2 Case Studies – Community Acceptance of Biosolids Projects

Beecher et al. (2004) also conducted in-depth analysis on 14 case studies of stakeholder interactions with biosolids projects (Figure 2-1). Examples were selected in order to demonstrate a variety of public acceptance issues, public acceptance strategies, and project outcomes. The case studies range from projects that created highly positive stakeholder relationships to projects that were stopped due to public protest. The case study findings allowed additional real-world perspective on subjects addressed by the literature review.

Public Acceptance Issues	1. King Co., WA	2. Milwaukee, WI	3. Jefferson Co., AL	4. Princeton, BC	5. Fulton Co., IL	6. Boulder, CO	7. New York City CAC	8. NYC and Rural Communities	9. Denver, CO	10. Philadelphia, PA	11. Montgomery Co., MD	12. Everett, WA	13. Kern Co., CA	14. New Hampshire
Social/Political														
Trust	•		•	•		•	•	•	•			•		•
Equity and Fairness	•		•	•	•	•		•	•	•	•	•	•	
Communications	•	•	•	•				•	•	•		•		•
Information (Sources and Timing)	•			•			•		•			•	•	•
Public Participation	•			•	•	•	•	•	•		•	•	•	•
Changing Demographics	•		•			•			•	•	•			•
Politics									•		•		•	•
Oversight and Enforcement	•							•	•				•	•
Scientific/Public Health														
Trace Metals and Chemicals	•	•						•	•	•		•	•	•
Pathogens								•		•	•	•	•	•
Odors and Other Air Quality				•	•		•			•				•
Water Quality			•	•					•	•		•		•
Soil and Food Quality										•			•	
Transportation/Trucking			•	•				•	•				•	•
Economic Viability			•		•		•	•	•		•	•	•	•
Emerging Issues and Uncertainty													•	•

Figure 2-1. Public Acceptance Issues Identified in the Biosolids Case Studies (Beecher et al., 2004).

2.4.2.3 Landowners' Perceptions of Biosolids Land Application

It is commonly understood that farmers and ranchers appreciate the benefits of biosolids land application. However, Krogmann et al. (2001) conducted interviews with 50 fruit and vegetable farmers in New Jersey (a state where land application is especially controversial) in order to better understand their perceptions and attitudes toward land application of sewage sludge. (Note: Authors refer to wastewater by-product as sewage sludge, not biosolids.) The research suggests that land application is not a common agricultural practice for these farmers. Twelve had considered land application, and seven had chosen not to participate after weighing benefits and risks. Thirty-eight had not even considered land application, and only one interviewee said he used “sewage sludge” on his land at the time of the interview.

While most of the farmers interviewed in Krogmann et al. acknowledged economic and agricultural benefits, they expressed concern about “too many unknowns” and about perceived risks, including heavy metals in sewage sludge (soil-build up, crop-uptake), negative public perception (and reduced consumer interest in their produce), odor complaints and contamination of the water supply (due to run-off). A third of those interviewed believed there were no benefits at all. A few voiced concerns about liability should consumers become ill, and many said they were somewhat to very uncomfortable eating crops grown in fields on which sewage sludge had been applied. Other concerns mentioned included decrease in land value, uneven application of sewage sludge, changes in regulations, municipal bans, and food processors’ perceptions of food products grown on land where biosolids had been applied. Over half commented that there was no increased health risk associated with land application. Many believed that the risks outweighed the economic incentives and soil improvement benefits. The researchers concluded that it was unlikely that these farmers would change their attitudes, or choose to participate in “sewage sludge” land application until their concerns were addressed and resolved. The authors recommended that the most effective way to communicate with farmers is through "direct field observation", such as demonstrations sites or neighboring farmers' applications.

2.4.2.4 Influences on Stakeholders' Judgments of Benefit and Risk

Empirical research dating back to the late 1970s by risk perception researchers, (Slovic, Fischhoff, and Lichtenstein, 1982; Flynn, Slovic and Mertz, 1994; Slovic, 2000) has shown consistently that a large number of factors can affect or influence people's judgments of the acceptability of risks. The key learnings from this research are directly applicable to understanding people's judgments on the acceptability of biosolids land application programs. Figure 2-2 shows eleven of those factors in descending order of impact.

Table 2-1. Influences on the Acceptability of Products or Technology (Thorne et al., 2006).

Less acceptable	More acceptable
Low trust	High trust
Benefits not clear	Benefits understood
Not controllable / Involuntary exposure	Individual has control
No alternatives	Voluntary exposure
Unfair distribution	Alternatives available
Dreaded consequence	Fair distribution
Affects children	Common consequence
Human origin	Affects everyone
High media concern	Natural origin
High symbolism	Low media concern
	Low symbolism

As Figure 2-2 shows, **stakeholders' trust in the people managing the risks can have the highest influence** on whether a person accepts a risk and the way it is being managed. People's judgment of trust involves many highly complex factors. For example, people's trust has been shown to be significantly influenced by whether they perceive: decision makers and risk managers to be competent and capable of appropriately managing risks and benefits; that decisions, actions and communications are open and transparent and balanced; and that they have meaningful input into the decisions that they believe can affect them.

Effectively responding to issues of trust requires that time and effort be extended toward developing relationships that promote shared understanding among program managers (and their employees) and community stakeholders.

The table also shows that:

- Benefits that are clear and well understood by individuals tend to lead to a judgment of the risk as being more acceptable than in situations where the benefits are unclear. When risks are perceived to be in the control of the individual, they tend to be judged as being more acceptable than risks that are perceived to be out of the control, or not very controllable, by the individuals, decision makers and stakeholders included. Risks where exposure is voluntary tend to be judged as more acceptable than risks where exposure is involuntary.
- Where distribution of benefits and risks, or gains and losses is deemed to be fair and balanced, risks can be judged to be more acceptable. If benefits and risks are deemed to be unfair in their distribution, so that one group is seen to experience or enjoy most of the benefits while most of the risks are transferred to or experienced by the other

group – the situation is deemed to be unfair and often leads to very controversial situations. This can often be the case in communities where biosolids are land applied; for example, the perception can be that the farmer gets all of the benefits – nutrients for his or her crops that result in better crop yield – and the near neighbors bear all the risks – the odor from the application.

- Dreaded consequences are much less acceptable than risks associated with more common consequences. This influence on risk perception might apply if people perceive a risk of catastrophic failure of a management option, such as a land application of biosolids that does not meet appropriate standards. Also, the most dreaded health consequence in North America is cancer. The real or perceived presence of cancer causing chemicals and other substances in biosolids may lead some to believe that the risks outweigh the benefits, regardless of the concentration of these substances in the biosolids.
- Risks that affect children are of great concern to people in virtually all cultures around the world. If children are not affected, or if the risk essentially affects everyone of every age, such risks are deemed to be more acceptable than those affecting children alone.
- Risks that are of natural origin (caused by "Mother Nature") are typically judged to be more acceptable than risks of human origin. Thus, stakeholders may come to different conclusions about the acceptability of risk, depending on whether they view biosolids as being of natural or human origin.
- Media coverage of risk can also affect people's perceptions and judgments of risk. If risks are given a great deal of media coverage – often with an intense focus on the nature of hazards rather than the actual level of risks (or benefits) involved – they may be judged by people to be less acceptable than those risks which are not the focus of media attention.
- Finally, risks that are highly symbolic, or evoke strong feelings in people, tend to be much less acceptable than risks not associated with any symbolism. Examples of symbols of identity in many communities may be wild animals (e.g., bald eagles), pristine natural landscapes and preserving a "rural way of life". Risks that threaten such areas can be deemed serious because, for many, strong feelings and personal identities are attached to these symbols for which no amounts of offsetting benefits may fully compensate.

Thinking about how these principles might apply to biosolids management, the researchers suggest people may perceive the risks associated with biosolids management to be higher if they:

- Do not trust the organization, the individuals, and/or their process for engaging stakeholders.
- Do not understand the benefits to them of a particular option, i.e., a biosolids land application in their community.
- Are not provided an ample opportunity to participate in decision-making processes.
- Believe that biosolids management options are being unfairly forced upon them.

- Perceive unfair distribution, particularly from rural community members who perceive that they are unfairly "burdened" with dealing with urban waste.

Again, in a biosolids management context, different perceptions of risk can be expected between those who clearly benefit from certain biosolids management options, and those who live, work, or play in areas that might be adversely affected by a biosolids management option. It may be as simple as people not wanting increased truck traffic, though they may present other arguments that they perceive to be more effective in meeting their interests. These kinds of sometimes hidden interests are more easily identified and responded to when communication processes are perceived to be fair, open, honest, and responsive to stakeholder needs.

2.5 Development of the Expert Model

From the insight generated through the literature review and through additional elicitation from the project team and other experts, the researchers developed the Expert Model, which presents a high-level diagram of the "system" of biosolids land application. The Expert Model was a critical tool to elicit and organize expert knowledge on the key issues in biosolids risk assessment (in particular, pathogen risk assessment), risk management and risk communication.

In August 2008, WERF hosted a joint meeting of the project team, WERF staff and the Research Challenge IAT. As part of this meeting, the researchers facilitated a discussion to gain insight needed to develop the Expert Model. Discussion topics included the influences on the need and opportunity for more effective biosolids land application risk assessment and risk communications processes; the range of stakeholders who may affect or be affected by biosolids land application; the range of processes for developing, storing, transporting and applying biosolids; biosolids pathogen risks; and potential impacts of biosolids land application.

From the input generated through the literature review and expert elicitation, the Simple Expert Model was refined, and the Detailed Expert Model drafted. The Expert Model was presented at the WERF Research Forum in December 2008, where WERF subscribers had an opportunity to comment. The Expert Model was also reviewed with each of the case study teams. Based on the feedback received, the Models were refined and finalized.

The Expert Model is presented in detail in Appendix A.

2.6 Preliminary Stakeholder Assessment

The first task in developing the Preliminary Stakeholder Assessment is to identify all of the stakeholders in risk management of biosolids land application, through *stakeholder mapping*. A stakeholder is defined as any individual, group or organization that may affect, be affected by, or perceive itself to be affected by decisions related to biosolids issues, projects, application, problems or policies. An understanding of the breadth of stakeholders – and their role in decisions related to biosolids land application and risk management – is important in understanding the influences on effective risk management and risk communications. A Stakeholder Map can be an effective tool in identifying the range of stakeholders who might have an interest in the decision making related to biosolids land application in their community, and in developing the Stakeholder Hypothesis.

To develop the broad Stakeholder Map, the researchers facilitated a workshop discussion among the experts on the WERF Challenge Research Team. Team members were asked to consider the following questions, focusing on biosolids land application, generally, and not site-specific:

- Which stakeholders would be the key decision makers related to biosolids land application and biosolids risk management?
- Which stakeholders would be most affected by the decisions regarding biosolids land application and risk management and have some sort of role or transaction in the decision-making process?
- Which stakeholders might be affected by biosolids land application and risk management, but are not directly involved in the decision-making process?
- Which stakeholders might have an interest in the topic and/or risks (impacts) associated with it, including health, environmental, social, technical, economic or political/regulatory, but are not directly affected by decisions?

Figure 2-3 presents the resulting *Stakeholder Map of Biosolids Land Application.* At the center of the Map is the risk challenge: public judgment of the acceptability of biosolids land application. Each of the groups or individuals identified has a relationship with a biosolids land application project that is more or less direct, depending on its proximity to the center of the concentric circles. Thus, the closer they are situated to the center, the more likely they will have a direct "stake" in decisions related to biosolids land application. The Stakeholder Map is described in more detail in Appendix A.

LEGEND

Decision Makers / Regulators: Stakeholders who have direct decision-making authority to establish rules about biosolids and/or can influence policy decisions about biosolids land application.

Transactors: Stakeholders who have some sort of role or direct transaction (e.g., financial) in the decision-making process.

Active Interests: Individuals or groups who have a stake in the issue of biosolids land applications and risk management (e.g., can be directly impacted), but are not directly involved in the decision-making process.

Audience: Individuals or groups that may have an interest in the topic of biosolids land applications and risk management, but who are not directly affected by nor involved in the decision-making process.

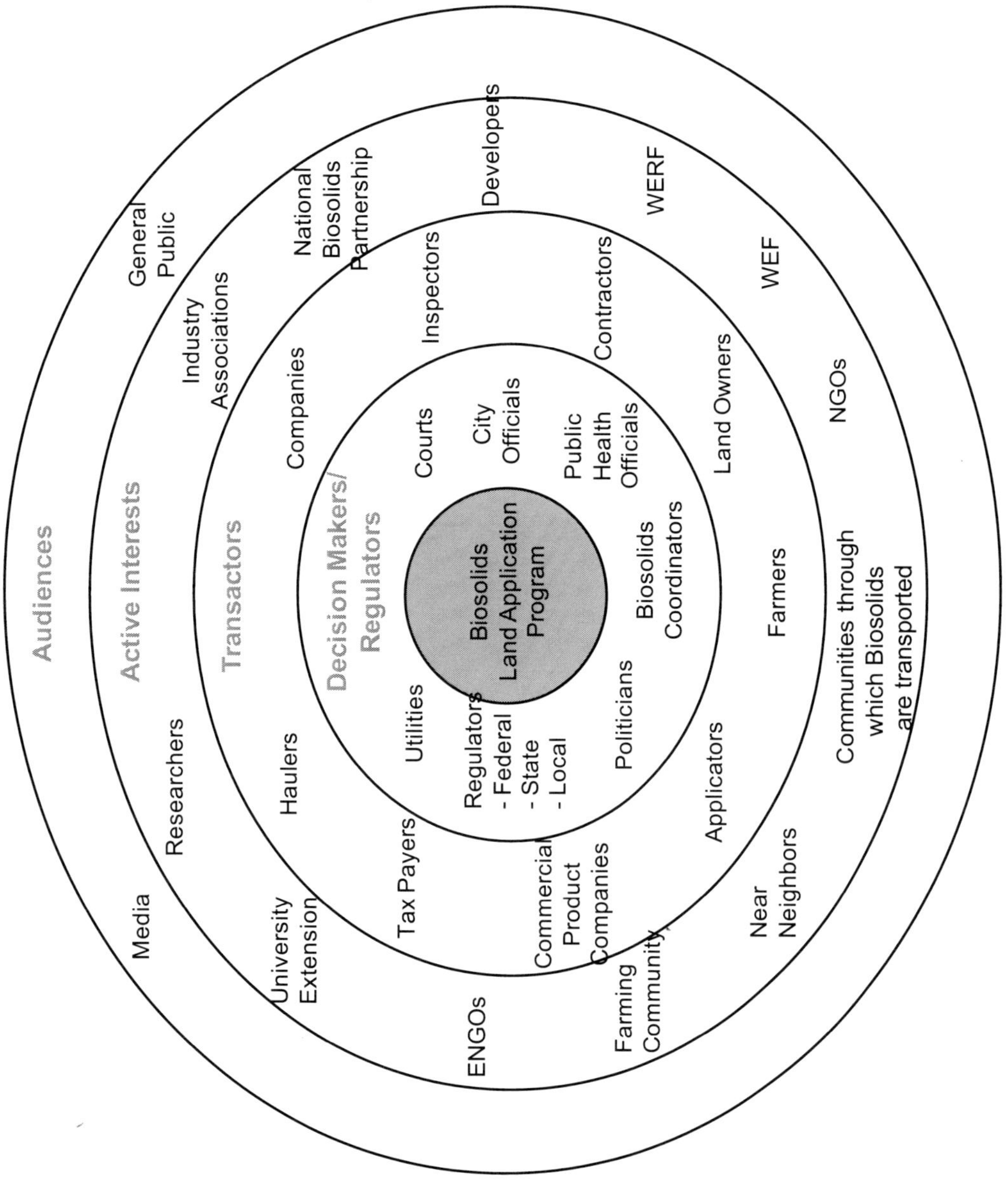

Figure 2-2. Stakeholder Map of Biosolids Land Application.

This map is intended to be generic, for the biosolids profession as a whole. In any given community, the map becomes dynamic and stakeholders can move into different categories depending on how they affect or are affected by decisions over time. For example, a landowner may first be Active Interests (*near neighbors)* and become Transactors (*landowners)* if they offer their land as a biosolids land application site.

2.6.1 Developing a Stakeholder Hypothesis

Next, the project team identified 'high priority' or key stakeholders, whose values, priorities, interests, and communications needs regarding biosolids land application and risk management are critical to successful biosolids land application. For the purposes of demonstrating the use of Stakeholder Hypothesis methods and tools, the following key stakeholders as highlighted:

1. **Near Neighbors:** These comprise near neighbors and community citizens who do not receive direct benefits; that is, they do not have a direct transaction from biosolids land application (e.g., payment or soil amendments), but who may otherwise be affected by land application and risk management decisions (e.g., bear a real or perceived risk, receive indirect benefits). Near neighbors are also key stakeholders because, while not necessarily influencing the decision-making process, their opposition to a site or applications can carry significant political weight.
2. **Applicators:** Haulers, contract land appliers, landowners and farmers or ranchers who are directly involved in transporting, storing and applying biosolids, and/or who make decisions regarding the use of treated land. They have a direct transactional relationship to a project or application in that they derive financial benefits from the construction of a "facility", transport of biosolids to and from the site in the form or raw material, and final finished products such as garden compost.

Using insight from experts and the literature review, a preliminary *Stakeholder Hypothesis* were developed for each stakeholder group. The Stakeholder Hypotheses summarize what is currently understood about the priorities, interests, perceptions, values and information needs of these key stakeholders. The following information was gathered to guide the development of the Stakeholder Hypotheses:

- What they know about the topic of biosolids and biosolids risk management.
- What they may not know or misunderstand that is consequential to decision making.
- Their values related to biosolids risk management.
- Who they trust to manage and communicate the risks of biosolids, and what factors they consider when making judgments or decisions related to biosolids risk management.
- What is the evidence of stakeholders' judgments or weighing of risks and benefits related to biosolids and biosolids risk management.
- What they *want* to know about the topic of biosolids risk management and what type of communications they prefer.

Based on a review of the case studies presented in Beecher et al. (2004), the researchers developed a Stakeholders Hypothesis for community neighbors. The key findings of that review were used to guide the design of the stakeholder research conducted as part of each case study (Chapter 3.0):

- Hypothesis: In communities where biosolids land application has been successful, community members (which would include landowners and near neighbors) generally understood the basic process of treating wastewater and converting the residual sewage sludge to biosolids. In general, this was not true in communities where biosolids land application has been unsuccessful.
- Hypothesis: In communities where biosolids land application has been successful, risk managers have actively engaged community members and worked with them to build understanding of biosolids and the opportunity it presents for beneficial use. Typically demonstration projects have been used during start up of the project. This approach has not been taken where biosolids land application has been unsuccessful.
- Hypothesis: Community members want specific details of a project, including where the waste is coming from, how environmental management will be conducted, how issues will be resolved and, generally, all aspects of project development. They do not trust information about applications at other locations, as the conditions for each application may be different, in terms of geography, weather, and community dynamics.
- Hypothesis: When making judgments of the acceptability of biosolids land application, community members consider many factors, such as location, public health issues, long-term impacts on food and water quality, quality of life impacts (e.g., odor), economic impacts, and opportunity for community involvement in decision making.
- Hypothesis: Community members place more trust in independent experts, independent facilitators, local government agencies, spokespeople from ENGOs, recreation and conservation groups, and public health officials to inform their judgments of the acceptability of biosolids land application projects.

CHAPTER 3.0

CONDUCTING RESEARCH WITH COMMUNITY STAKEHOLDERS

3.1 Overview

As a key component of the case study applications of the Strategic Risk Communications Process (Step 3), the researchers conducted in-depth mental models research with a sample of 48 individuals in four communities (Tulsa, Oklahoma; Lunenberg, Tidewater and Fauquier County, Virginia). An additional 6 interviews were conducted with Virginia Department of Health officials as part of the Virginia Case Study. Detailed summaries of the research are provided in each of the Case Study Executive Summaries (Appendix B and Appendix D).

Note: The researchers recognize that many biosolids professionals may not have the financial or research capacity to conduct formal research using the mental models research methodology in their own communities. The researchers believe that there is general insight that can be taken from the learnings of the two case studies that can be useful to biosolids professionals in other communities. While the case study findings are specific to Tulsa, Oklahoma, and Lunenburg, Tidewater and Fauquier County, Virginia, the general learnings may apply to other communities. It is important that biosolids professionals do at least some informal research of their own on the Strategic Risk Communications Process. Depending on the circumstances of the biosolids program and available resources, conducting mental models research as part of the process may be warranted. Any communications should be pre-tested in the local community to ensure that they are interpreted as intended.

Based on the literature review and the Stakeholder Hypotheses, the following cohorts were identified as key stakeholder groups, whose values, priorities, interests, and informational needs regarding biosolids land application and risk management are critical to the success of sustainable biosolids land application and, in each of the case studies, successful business. The research sample for this project comprised:

- Near Neighbors: Individuals who reside in or own property within approximately one mile of a current or potential biosolids land application site (Tulsa and Virginia). Near neighbors and other community members may not receive direct benefits from biosolids land use (e.g., payment or soil amendments), but may otherwise be affected by land application and risk management decisions (e.g., bear a real or perceived risk and/or receive indirect benefits).
- Landowners (Tulsa only): Individuals who offer their property for biosolids land application. Landowners who choose to apply biosolids on their land may receive direct benefits and may, therefore, have a different stake in projects. The case study teams saw landowners as biosolids communication partners, playing a potentially important role in communicating with neighbors and other community stakeholders. Also included in this cohort were potential landowners – farmers and ranchers who had expressed interest in biosolids land application but had not yet received them.

- VA Department of Health (VDH) officials (Virginia case study only). These are individuals who work for the VDH and who would view biosolids safety as a top priority. VDH officials may be a direct source of information for people who inquire about biosolids safety.

The objectives of this research were to: a) elicit the values, priorities, interests and informational needs related to biosolids land application of these key stakeholders in two geographic areas – Oklahoma and Virginia; b) support development of risk communications tools and core messages that can be adapted by biosolids professionals; and c) use the results of the case studies to develop prototypical communications materials. Another objective was to inform the technical pathogen risk analyses and tool development by the Biosolids Pathogen Risk Assessment Team.

3.2 Research Preparation

The research followed an Expert Model/Mental Models research approach (Morgan et al., 2002), described in Chapter 1.0. The researchers worked with the project team and each case study team to develop a research plan, tailoring the plan to fit the context of the biosolids land application programs and the community(ies). The researchers ensured that the appropriate steps were taken to address specific issues and sensitivities of the host utilities, applicators and communities. The mental models research process is a very respectful one for all stakeholders, including the research sponsors and case study participants. Experience has shown that even interviewees who are critical of the profession or technology will often express satisfaction and gratitude for having the opportunity to express themselves in a meaningful way and have their interests heard and recorded by the researchers.

3.2.1 Interview Protocol

A semi-structured interview protocol (Appendix E) was developed to guide the research interviews. The protocol was designed to solicit interviewees' thinking about the topics defined in the expert model (See Chapter 2.0 and Appendix A). It was developed based on the key learnings from the Stakeholder Hypotheses and in consultation with the project team. It was refined as necessary for the case study research applications, in consultation with each case study team.

The interview protocol offered interviewees an agenda of topics to discuss in a way that allowed for free expression. Key topics included:

Part 1: General thinking on biosolids: These questions addressed what landowners and near neighbors know and/or believe about biosolids and their use, including:

- Familiarity with and top-of-mind thinking on the terms "biosolids" and "sewage sludge".
- What, if anything, they have heard or learned about biosolids and through what channels. Note: At this point in the interview, the researcher provided the interviewee with descriptions of 'biosolids' and 'biosolids land application', in order to facilitate a shared understanding between the interviewer and the interviewee and also to facilitate a common basis for comparing the mental models and expert models.
- Thoughts on the reasons why biosolids may be produced.
- Perceptions of the process of converting wastewater to biosolids.

- Perceptions of biosolids land application and who is involved.
- The interviewee's experience with biosolids land application.
- Perceptions of what happens to land over time after biosolids have been applied.

Part 2: Benefits and risks associated with biosolids land application: These questions addressed landowners' and near neighbors' perceptions of the benefits and risks associated with biosolids in land application, with particular attention paid to pathogens, including:

- Thoughts about the potential good impacts (benefits) associated with biosolids land application.
- Thoughts about potential negative impacts (risks) associated with biosolids land application, and factors that increase or decrease the potential risk.
- Thoughts on pathogen risks and their potential impacts, including pathogens from other types of fertilizers.
- Comparisons of benefits and risks of between biosolids and other types of fertilizers.

Part 3: Decision-making process for biosolids land application: These questions addressed landowners' and near neighbors' perceptions of the decision-making and risk management/communications processes related to biosolids land application in their communities, including:

- Perceptions of who are the 'key players' when it comes to making decisions about land application biosolids projects in their area, and their trust and confidence in each of these key players to effectively manage the risks and benefits of biosolids application.
- Thoughts about the factors key players should consider when deciding whether to use biosolids in land application.
- Perceptions of government regulations related to biosolids.
- Overall judgments (at this time) of the acceptability of using biosolids in land application in their community. What factors would affect that support?

Part 4: Communications related to biosolids land application: The questions in this section focused on landowners' and near neighbors' perceptions of the need for communications related to biosolids land application in their communities, with particular attention paid to communications related to pathogen risk management. Questions focused on:

- Questions that the interviewee would like to have answered about biosolids.
- Preferred and trusted communication sources (*who* is communicating) and modes (e.g., Internet, meetings).
- Thoughts on how biosolids land appliers could best communicate with people in the future.

Two pre-tests were conducted (not analyzed) to ensure the protocol questions were interpreted as intended. The information collected from these interviews was used to refine the protocol and prepare the analysis.

A similar protocol was developed for the VA Department of Health officials, but focused more on their professional roles and responsibilities related to biosolids land application.

3.2.2 Research Participant Sampling and Recruitment

The research samples were generated using two methods. In Tulsa, a list of current and past landowners who receive land-applied biosolids was provided by the case study team. A list of people who have expressed interest in biosolids land application to the City of Tulsa Public Works Department (referred to as potential landowners) was also provided. For the VA research with VDH officials, a list of regional Department of Health officials was provided by the case study team. Participants were chosen at random and recruited to participate until the targets sample sizes were met.

The neighbor samples were generated in both case studies through Internet-based mapping and reverse address look-up. The case study teams provided the location's address or intersections, or geographic coordinates of current or recent land application sites. The researchers identified the field on Google Maps® and identified streets within a specified distance from the application site, in any direction.Yellowpages.com® was used to identify residents (with active land-line phone numbers) along those locations. Contact information from Yellowpages.com was included in a database. Researchers chose randomly from the participant list, ensuring that there were no more than two neighbors recruited who resided in very close proximity (e.g., "next door") to a single application site. Participants were chosen at random and recruited to participate until the targets sample sizes were met.

Table 3-1 lists the number of potential participants who the researchers attempted to contact in each sample, as well as the number who accepted the invitation to participate and the number who declined the invitation to participate. Percents do not total 100% because the table does not include the number of potentials contacts for whom contact was attempted but no contact was made (e.g., no answer, bad numbers).

Table 3-1. Summary of Response Rates for Each Cohort.

		Tulsa Case Study	VA Case Study
Neighbors	Attempted contact Accepted Declined	149 12 (8%) 68 (45%)	321 24 (7%) 95 (29%)
Landowners	Attempted contact Accepted Declined Did not quality	30 12 (40%) 12 (40%) 1 (3%)	NA
VDH	Attempted Contact Accepted Declined	NA	18 6 (33%) 6 (33%)

Eligible participants included adults over the age of 18. Where possible, researchers targeted to achieve a mix of participants with regard to age and family status.

3.2.3 Conducting the Interviews

Interviews were scheduled and conducted by interviewers on the research team who were trained and oriented to this project. Researchers contacted potential participants by telephone, inviting them to participate in the research. Potential interviewees were assured that their comments would be confidential to the research team. Interviewees were notified during the invitation that they would be offered a $30 honorarium for participating. If the potential interviewee agreed to participate, the researcher scheduled a phone interview at the person's convenience.

All interviews were conducted by telephone, with researchers following the interview protocol. Interviewees were encouraged to respond to questions as they understood them. They were also encouraged to raise additional topics spontaneously and to elaborate on their perspectives. Interviews were audio-recorded with the permission of the participants and near-verbatim transcriptions developed. To preserve confidentiality, names and other identifying information were not included in the transcripts.

3.2.4 Analysis and Reporting

In the analysis phase, each interviewee's responses from the transcripts were parsed into discrete data segments. Each segment was assigned one or more codes that best aligned with the various nodes in the expert model. New codes were generated in cases where the model nodes did not sufficiently capture the theme as it was expressed in the segment. Descriptive keywords were also assigned to capture finer details related to a particular node. Coded segments were considered "spontaneous" mentions if the interviewee's response was generated without prompting the interviewee to think of a particular topic. Segments were coded without considering the accuracy of the interviewee's statement.

Initially, several interviews were coded independently by two individuals. Differences were resolved by discussion. Once the coding procedure seemed stable, a single individual coded subsequent interviews. Coded interviews were transferred into a database where more detailed analyses were conducted on segments by interview question, code, keywords, or in combination in order to subjectively assess the results captured within each node and to identify other emerging themes. The coded results were organized into mental models diagrams, in a format aligned with the expert model.

Separate analyses were conducted for each of the research cohorts in order to assess potential variations between the cohorts. Additionally, the results were compared against the stakeholder hypotheses generated by each case study team in order to assess what these stakeholders know that is correct and what they do not know or misunderstand that is consequential. The gaps and alignments were then used to develop prototypical communications plans, messages and materials.

3.3 Overview of Key Research Findings

Mental models interviews were conducted with: 12 landowners and 12 neighbors in Tulsa, Oklahoma; 24 neighbors in three Virginia communities (Fauquier County, Tidewater area and Lunenburg); and 6 Virginia Department of Health officials. Detailed research findings and a discussion on how they were incorporated into each case study communications plan are included in Appendix B, Tulsa Case Study Application of the Strategic Risk Communication Process, and Appendix D, Virginia Case Study Application of the Strategic Risk Communications Process.

This section provides a highlight of the general research findings across all research cohorts. These research results can provide biosolids professionals some insight into the range of priorities and perceptions that may be shared by people in their own communities. They also provide in-depth insight that may provide additional context to prior research on public perceptions of biosolids land applications (e.g., Beecher et al., 2004; Krogmann et al., 2001; Robinson and Robinson, 2006.) A key difference between this research and much of the prior research is the focus is on specific populations, particularly near neighbors to land application sites, rather than a general population. (Note: These qualitative, small-sample research results should not be interpreted as being representative of the U.S. population. Use of percentages are used only to indicate the proportion of the sample that responded a particular way in the research.)

Sample quotes are included to provide context for the findings and to provide deeper insight into interviewees' thinking, in their own words.

3.3.1 General Thinking on Biosolids

At the beginning of the interview, interviewees were asked what first comes to mind when they think about "biosolids". A variety of mental models regarding biosolids and biosolids land application emerged. Interviewees' perceptions of biosolids ranged from "I would just like as much as I can possibly get" (a Tulsa landowner) to "a low-grade fertilizer" to "human waste and contamination" to "I'm not sure exactly what it is."

Most Tulsa landowners and neighbors demonstrated general awareness of biosolids as the by-products of wastewater treatment. A few specifically mentioned that these by-products were treated before being land applied.

- "It's the biological active material that's left over from wastewater treatment." (*Tulsa landowner*)
- "It's the solid material left over after treatment and pumping off of the water, which mostly would be human waste." (*Tulsa landowner*)
- "Biosolids are solid waste that has been treated and put back into the environment in many different fashions." (*Tulsa neighbor*)

Interviewees in VA expressed a wider range of familiarity with biosolids. A few interviewees, all from Fauquier County, offered accurate descriptions of biosolids before any descriptions were provided by the interviewer. (Descriptions were considered "accurate" if they included statements about wastewater by-products (e.g., human waste) and either treatment or re-use.) For example:

- "The by-product or the end result product of the wastewater treatment process is the solid part to absorb the chemicals and it's super heated and a lot of the germs and diseases are killed in that process and the products are used for application on farm land for fertilizing purposes." (*VA neighbor*)

Most VA neighbors, however, were much less familiar, and offered a range of perceptions (before the descriptions were provided), including: "something from the earth that produced a solid", "the living things that break up the gunk in waste treatment", "pet waste", and "sediment that settles in the bottom of a tank".

Some interviewees, in all cohorts, demonstrated specific knowledge of wastewater and/or biosolids treatment. However, it appears that many interviewees have very little understanding of wastewater or biosolids treatment processes. Many commented that wastewater is treated with chemicals, but they were unfamiliar with what chemicals were used. A few said that they did not know, or did not believe, that wastewater is treated (e.g., "It is pumped into the Potomac River in DC." *VA neighbor*).

More broadly, interviewees, particularly neighbors in Tulsa and VA, generally did not appear to have an in-depth understanding of water and wastewater systems. Some commented that they were "not a part" of the wastewater system. A few said that because of this they don't pay attention to wastewater issues. Others said that they are burdened – taxes, disposal – by water system that they "are not a part of":

- "It's not really a subject that I focus on because of where I live. I live out in the country, so I have a well and a septic field, so municipal wastewater is not really something that's at the top of my mind." (*VA neighbor*)
- "I'm getting taxed for something that I don't necessarily have." (*VA neighbor*)

Nearly all Tulsa interviewees (including most Tulsa neighbors) were aware that biosolids are land applied in their community. Most landowners demonstrated specific knowledge of the biosolids land application process and described various pre- and post-application requirements. Most Tulsa neighbors demonstrated basic knowledge of land application (e.g., spreading on fields), and commented on direct experiences seeing (and/or smelling) land application.

- "I've seen the truck out there. They have these scratchers that go along beside the truck or beside the big tractor similar to a combine harvester and they squirt it out onto the ground or inject it into the ground." (*Tulsa neighbor*)

Most Tulsa landowners appeared to be very familiar with pre- and post-application requirements. They described specific rules and expectations they are expected to follow. However, a few appeared to be less familiar (and suggested possible lack of full compliance) with site restrictions and other requirements.

Most VA neighbors from Fauquier County were familiar with the local land applications, and a few commented that biosolids is a "controversial topic" in the area. A few commented on having had a negative experience with biosolids (e.g., seeing a spill that had to be cleaned up). Many neighbors from Lunenberg County said they were familiar with biosolids. Interviewees in Tidewater were generally not familiar with biosolids and none had direct experience with biosolids land application (e.g., "I haven't paid attention"); however, some speculated that biosolids may be applied on fields near their homes (e.g., by smelling an odor, seeing something sprayed, etc.).

When commenting on biosolids, VA neighbors, particularly those who appeared to be less familiar with biosolids, often drew on their perceptions of other environmental and health issues – and how these issues were managed by the people in charge. Examples included: the Gulf oil spill; PCB contamination in the Chesapeake Bay; fly ash and coal ash contamination on public land; animal manure; and commercial fertilizers.

3.3.2 Interviewees' Judgments on Local Biosolids Land Application

Throughout the interviews, interviewees thoughtfully considered the perceived benefits, risks and cost tradeoffs of biosolids land application in their communities. Interviewees believed that the most important considerations in decisions regarding biosolids land application are the quality and oversight of regulations, the safety of biosolids, and the impact on neighbors and the community.

Near the end of the interview, interviewees were asked if they were *generally in favor* of biosolids land applications in their county, *generally not in favor* of biosolids land applications in their county, or undecided. Figure 3-1 provides an overall summary of the responses across all interviewee cohorts:

- All Tulsa landowners were in favor of biosolids land application.
- Tulsa neighbors were generally not in favor, citing the odor, health concerns, quality of life impacts, and uncertainty. A few neighbors commented that they were generally in favor of biosolids, under certain conditions, such as "if they are controlled, regulated, tested and a 100% safe" and "if it is done properly".
- 50-60% of the VA interviewees in each community said they were "in favor" of biosolids. They cited the value to the agricultural community; responsible management of a societal problem; and confidence in biosolids management. About 10% were not in favor, citing the unacceptable uncertainty about health and environmental risks. The remainder (about 30%) said they were undecided or in favor "with conditions". They wanted more assurance about the safety and long-term health and environmental impacts.
- VDH officials commented that they were in favor, generally referred to biosolids as the "best of both worlds".

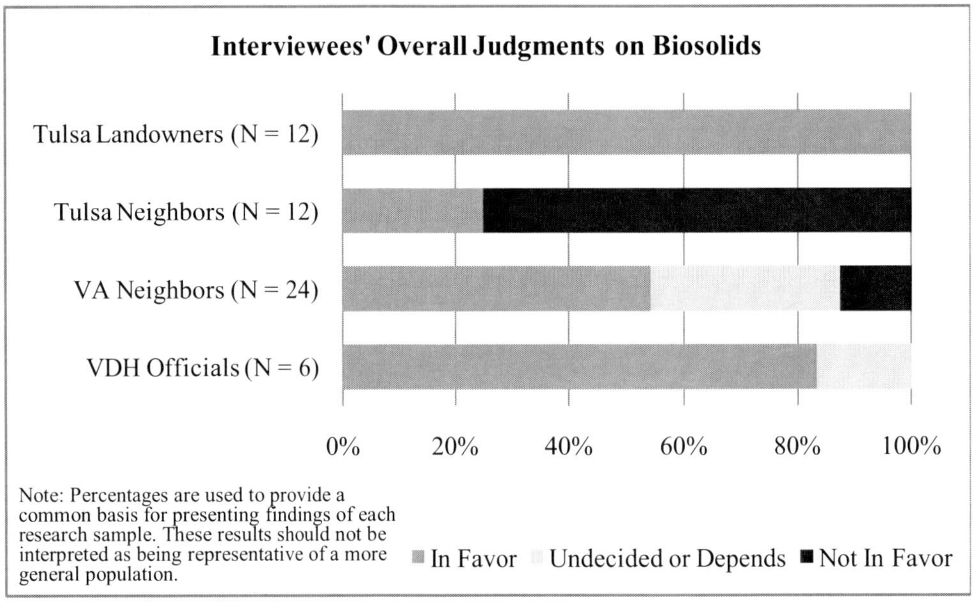

Figure 3-1. Summary of Judgment Response for Each Research Cohort.

Most interviewees provided thoughtful comments on the weighing of tradeoffs between the benefits and risks of biosolids. For example, a few who said they were "in favor" of biosolids commented that while there is a smell, it is temporary and an expected "part of farming". A few spontaneously compared the benefits and risks of biosolids to the benefits and risks of using animal manure or chemical fertilizers on land, noting that biosolids are not "any more toxic or dangerous". However, all of those who said they were not in favor of biosolids appeared to have thoughtfully considered the risks and benefits and concluded that the risks do not outweigh the benefits. For example:

- "Opponents of sewage sludge as fertilizer say that there have been studies that say it cannot be thoroughly treated. I think anything can be properly treated. My only concern would be that it be treated in an ecologically sound matter. I don't think treating sewage sludge with hazardous chemicals to kill bacteria may be the answer." (*VA neighbor*)

- "It does smell when they spread the stuff, but the smell only lasts a couple of days. If you're in a rural situation, you realize that stink is a part of farming, as opposed to someone who builds a million dollar house out in the middle of nowhere and then wonders why it smells sometimes." (*VA neighbor*)

3.3.2 Influences on Interviewees' Judgments

The following is a summary of the factors that appear to influence most strongly (positively or negatively) interviewees' judgments on the acceptability of biosolids land application in their counties. They are listed in an approximate order of "strength" of influence, as determined by a qualitative summary analysis across all cohorts:

<u>Individual's awareness and experience with biosolids.</u> Those interviewees, particularly Tulsa landowners and VDH officials, who appeared to be more familiar with biosolids land application tended to say that they were "in favor". Some neighbors said they felt "comfortable" with biosolids, because they grew up in an agricultural environment.

- "I was raised to know what sludge was so I have always known about it and felt comfortable with it." (*VA neighbor*)

<u>Confidence in the safety of biosolids.</u> Interviewees who were in favor of biosolids tended to express confidence (e.g., "trust") that biosolids are safe. Tulsa landowners expressed confidence in the safety of biosolids land application, citing their trust in the regulations and the City of Tulsa and the "naturalness" of biosolids versus "man-made" products. VDH officials also generally expressed confidence that "when all the rules are followed, it appears to be a safe procedure". The few Tulsa neighbors who expressed confidence in the safety of biosolids commented that they had "no concerns" and thought "[the people who are applying it] were doing the right thing".

Neighbors in Tulsa and VA who were "against" or "undecided" expressed less confidence in the safety of biosolids land application, citing a lack of confidence in the decision makers and the regulations, and the uncertainty and "newness" of biosolids. These interviewees also talked about "not seeing enough data" or "being ignorant" of the risks to be able to conclude that biosolids are safe.

<u>Odor.</u> Almost all Tulsa interviewees attribute odor to biosolids land application. Tulsa landowners believed odor was a less significant concern because it is "minor", "short-lived" and "worth it". Tulsa landowners were familiar with their neighbors' complaints about odor. As one commented, "If their nose is a little delicate, but this goes away in a few days and now it doesn't smell at all."

Tulsa neighbors cited odor as a significant concern and attributed odor to health, quality of life and livestock impacts. Odor was the most prominent factor cited by those Tulsa neighbors who said they were "against" biosolids land applications in their county:

- "I've seen them spreading it and smelled to the point where it just sickens you. Actually, I can tell I've got a respiratory problem right now. It came on about the same time the stench started." (*Tulsa neighbor*)
- "There was a strong odor and the flies were terrible. I don't know if they multiplied because of it but if you went to [one landowner's] house, you couldn't leave the windows down on your pickup or it would be completely full of flies." (*Tulsa neighbor*)
- "I have to actually cancel people who might visit me because it smells so bad." (*Tulsa neighbor*)
- "I even heard from ranchers that their cows will not eat that grass because of that stench bringing into the roots and going up in to the grass." (*Tulsa neighbor*)

Odor did not appear to be as significant an issue for VA neighbors generally, with the exception of some interviewees in one VA community who commented on the impact of the odor on their quality of life and property values.

<u>Concerns about contamination,</u> which interviewees believed could result in a number of negative health and environmental impacts. The most prominent concerns about water contamination, particularly well water. "Run-off" into residential areas was also raised as a concern. Other concerns included pollution of streams, air pollution.

- "The first hard rain you have there is going to be run-off and the run-off is going to have biosolids or whatever they have fertilized their fields with, which will go directly into the water source which are the streams or ponds nearby or go directly into the ground and contaminate your well." (*VA neighbor*)
- "People live across the street from this thing and that water is running right off onto their front yard where babies are playing." (*Tulsa neighbor*)

Some interviewees, in the Tulsa landowner, Tulsa neighbor and VA neighbor cohorts compared biosolids to chemical or animal fertilizers that are applied on the local fields. These interviewees generally expressed less concern about biosolids, relating biosolids as a "natural product":

- "The bigger problem is the cattle, pig and chicken waste coming out of Arkansas into the river." (*Tulsa neighbor*)
- "[Biosolids are] somewhat natural and the other fertilizers we know have harmful chemicals in them. One of the issues we have is keeping the harmful chemicals out of the ground and from that standpoint, that's the plus, at least to me, to use [biosolids because they are] less likely to be harmful to the land." (*Tulsa landowner*)
- "Chemical fertilizers leach into the watershed a lot more effluently than biosludge and have a higher risk of toxicity and/or proven carcinogens." (*Tulsa neighbor*)

<u>Concerns about the "unknowns" about biosolids</u> (e.g., emerging contaminants and long-term impacts). In particular VA neighbors raised concerns about long-term impacts. VDH officials also expressed concerns about emerging contaminants and the "uncertainty of the science", and focused on the long-term impacts and "non-infectious compounds", such as hormones, medications, heavy metals and "volatile organic compounds":

- "If you pump something into the ground for however many years, we don't know what it's doing. This is a fairly new thing... but in a hundred years, who knows? You know, nobody really knows. That's a major negative impact." (*Tulsa neighbor*)

Trust in the people who produce and apply biosolids. This included interviewees' confidence in the adequacy of the regulations to protect against potential health and environmental impacts. Most neighbors commented on their lack of trust in decision makers, some spontaneously mentioning that they do not trust the government and/or city officials. Some Neighbors mentioned that they do not trust the utility citing motivation, regulations and site selection as the key reasons for their lack of trust.

Agricultural and economic benefits to farmers and to the community. Interviewees who said they were "in favor" of biosolids land applications in their counties emphasized their economic benefits. Tulsa landowners cited (and in some cases quantified) numerous direct benefits including richer soil, faster growth of cattle and increased crop production, cost savings on fertilizer and increased revenue. Neighbors also commented on the benefits – speaking to them more as indirect rather than direct benefits of biosolids. Many believed that biosolids are low-cost or free, and that "landowners want it so it must have some benefits as a fertilizer."

A few neighbors questioned the agricultural benefits of biosolids:

- "They say it makes the soil nutrient rich and it helps the crops to grow. It's basically excrement, so how is that nutrient rich? When animals digest things, a lot of the nutrients are still in their excrement. With humans, usually the body takes the nutrients out … I've talked to farmers and nobody has ever been able to answer that question for me." (*VA neighbor*)

Other benefits cited included economic benefits to the utility (a way to "get rid" of "it" and "lower cost than dumping"), environmental benefits and economic benefits to the community. One interviewee commenting on the economic benefits to the community noted, "If they did it right, they could reduce everybody's bill by selling biosolids to the farmers and thereby subsidizing the wastewater treatment plants themselves."

Both landowners and neighbors in all cohorts, mentioned environmental and economic benefits to the community, commenting that biosolids land application is a low-cost and environmentally friendly way to manage ("get rid" of) a wastewater by-product, which benefits the community as well. When commenting on the benefits of biosolids, VDH officials generally referred to them in terms of the "best of both worlds" pointing out the agricultural benefits of biosolids and their role in waste management.

Other factors affecting judgment included: the perceived source of biosolids (a few interviewees were in favor of biosolids that come from the local communities); perceived use of chemicals in treatment of biosolids (a negative influence); perceived risk of hazardous waste (a negative influence), perceived risk of over-fertilization, and impacts to wildlife. For example:

- "The majority of time when human sludge is put on a field you will not see wildlife on the field because the smell, the scent will draw them away. The wildlife itself, especially deer and turkeys around here, are not going to graze on that field because of the smell." (*VA neighbor*)

3.3.3 Safety of Biosolids, Including Pathogens

Interviewees were read a description of pathogens in biosolids and removal during treatment and then asked to provide comments and questions about pathogens. Note: There was very little spontaneous mention by interviewees in any cohort (with the exception of a few VA neighbors) of pathogens before this description was read.

After hearing the description, many interviewees appeared to be generally familiar with pathogens. *E. coli* was most often mentioned. Anthrax and salmonella were also mentioned. They talked in vague terms about the health impacts of pathogen exposure (e.g., get sick). Their comments primarily drew on previous events or experiences (e.g., a previous outbreak, suspected contamination incident, or a recent food recall).

Nearly all Tulsa landowners expressed a low level of concern about pathogen exposure, citing their trust in the utility and the treatment of the biosolids, their perceived risk of other agricultural sources, and their personal experience using biosolids:

- "We've noticed nothing on our cattle." (*Tulsa landowner*)
- "Someone has given some thought to this and decided that is the level that they believe is safe out there [and has a] safety margin." (*Tulsa landowner*)
- "I don't ever worry too much about unsafe levels. I was raised around here with our drinking water out of a pond run through a simple filter. And I raised four of my own and 25 or 30 of other people's and they all survived." (*Tulsa landowner*)
- "You don't have to have biosolids on it to have problems with pathogens. The cows can have disease. *E. coli* is naturally in the soil." (*Tulsa landowner*)

One Tulsa landowner, however, expressed concern, recalling a direct experience: "We had an *E. coli* outbreak that may have been caused by poultry manure application and so there's a real heightened interest in this subject."

Some neighbors, both VA and Tulsa, also expressed less concern about pathogen exposure. When asked and how they would compare the risks of biosolids to the risk of the other sources of pathogens, many VA neighbors indicated they believed the risks were the same: One Tidewater interviewee commented, "a pathogen is a pathogen", another VA neighbor commented, "I know [grazing] animals are already exposed to bacteria in manure. It is obvious that these materials do not last long-term or we would have destroyed ourselves a long time ago."

However, most Tulsa neighbors, expressed a higher level of concern about pathogen exposure upon reading the description. A few VA neighbors indicated they are more concerned, citing the unknowns about biosolids. If you don't reduce the pathogens you can catch some very nasty things. There are some instances of people dying from biosludge. (*Tulsa neighbor*)

- "You never know if bacterial gets through that it will start breeding if it is not absorbed into the soil with enough rain, if it is not cleaned of bacteria that it builds up and we all get sick." (*Tulsa neighbor)*
- "Human sewage carries a lot of things, I just don't know if even with treatment it can be absorbed by human consumptions, vegetables, or absorbed by animals grazing on the land." (*VA neighbor*)

VDH officials appeared to be comfortable with the safety of biosolids in relation to pathogen risks. Some commented that with regard to pathogens, biosolids land applications has "lots of scrutiny", "proven long-term use", "great deal of experience", "no documented problems", and the "ability to "track" or identify problems. They were more concerned about emerging contaminants and the "uncertainty of the science". They also focused on the long-term impacts and "non-infectious compounds". They expressed some concerns about researchers', regulators' and biosolids managers' understanding of the complexity of the long-term health and environmental impacts of biosolids use", such as hormones, medications, heavy metals and "volatile organic compounds".

3.3.4 Confidence in Regulations and Decision Makers

In the interview, interviewees were read several descriptions of regulations followed by the biosolids professionals in the various communities. A prevalent theme among interviewees in all cohorts was the adequacy of regulations in place, and whether the regulations are being followed. When read a description about regulations regarding when, where and how biosolids can be applied, many commented that they wanted to know *why* the regulations are in place – that is, whether the regulation indicates some sort of risk:

- "What are the hazards that you need to stay away from the land for at least 30 days and why does it need to be away from homes and water supplies?" (*VA neighbor*)
- "I understand that they're going by the governmental guidelines, but that leads me to ask how are you so sure that the governmental guidelines are good enough?" (*VA neighbor*)

Some neighbors raised questions about the adequacy of the regulations:

- "The odor issue involved is very egregious to the surrounding area, 250 feet doesn't mean anything to a wind." (*Tulsa neighbor*)
- "Rules and regulations – where it can be applied, how long for, how not and if they are 100% UFTA or government approval. That is like throwing tissue up in the air. It has got to be more controlled. And more guaranteed and more secure." (*Tulsa neighbor*)
- "100 feet from the water supply does not mean it is not going to get into the water supply, just look at the rain we have had today." (*Tulsa neighbor)*
- "The land in this area has drainage ditches and the water runs off to the river when it rains. My property is 800 feet long so keeping this material a few hundred feet away doesn't seem like a long way." (*VA neighbor*)
- "100 feet is not far enough if being sprayed given previous knowledge of what happens with chickens and water ways." (*VA neighbor*)

Tulsa landowners, VDH officials and some neighbors (primarily in VA) expressed confidence in the regulations:

- "I think for sure that it can be done safely and with the regulations and the requirements imposed on the application of it that the land can absorb it safely without having the adverse effects on humans and the environment. (*VA neighbor*)

- "If the government is doing something like that, I guess they have to know what they're doing."(*VA neighbor*)
- I would think that the treatment plants do what they're supposed to do to make sure they are safe." (*Tulsa landowner*)

Most neighbors in Tulsa spontaneously expressed concern about whether the regulations were being – or could be – followed:

- "People aren't waiting thirty days and how far it is away from people? That's very important. They need to block the road off."
- "250 feet by my guess ... How wide do you think a street is? Maybe two cars wide, so we're talking not 20 feet. Well, the fence is on the edge of the street. All you've got is a ditch, and then people live across the road. There's no way that's 250 feet."
- "I have no idea if the City of Tulsa does soil samples as they are required to do and checks for over application."

Discussions related to the level of confidence in decision makers elicited a wide variety of responses among interviewees. Nearly all Tulsa landowners expressed trust in biosolids decision makers generally speaking and most expressed trust in the utility. Most Tulsa neighbors commented on their lack of trust in decision makers: some spontaneously mentioned their lack of trust in the government and/or city officials, others mentioned their lack of trust in utility, citing motivation, regulations and site selection as the key reasons for their lack of trust. VA neighbors, however, demonstrated a lack of understanding of who is "in charge" of biosolids land application and spoke in very general terms about decision makers. Throughout the interview, VA interviewees offered their judgments of the credibility (both positive and negative) of the people they thought may be key decision makers – the local governments, utilities, biosolids companies, regulators, agricultural extension officials and farmers.

3.3.5 Communications Related to Biosolids Land Application

Many neighbors, both in Tulsa and in the VA communities, spontaneously commented on the lack of communications with the utilities, organizations and companies involved with biosolids. VA interviewees, in particular, commented that they have not received information or engaged in direct conversation with the people who manage and use biosolids in their communities. They specifically mentioned that there was no "advance notice".

Interviewees commented on several opportunities to improve the quality of engagement with people in their communities about biosolids and to increase the acceptability of biosolids land application. VDH officials, in particular, stressed the importance of community acceptance of biosolids, commenting that they believed that "People need to be sold on it." Opportunities for improving community engagement included having more transparency in decision making; communicating in advance of a land application; providing information on biosolids treatment and applications processes; providing balanced information on the benefits and risks; and providing evidence and/or an assurance of safety. VDH officials suggested additional information also should be communicated, including information about past and ongoing research; the value of biosolids; and the societal responsibility to manage wastewater by-products.

When asked *who* should communicate about biosolids, the most common response was government officials, particularly federal and state agencies such as the EPA, DEQ, or local agricultural extension agents. These officials were believed to have particular expertise and less vested interest. Communications from the utilities, or those within the biosolids profession, elicited mixed responses. While many Tulsa landowners commented positively about their engagement with the City of Tulsa Public Works Department, stating that the Utility had been "fair", "prompt" and "forthcoming with information", and a few VA neighbors mentioned the "biosolids people" (e.g., utilities and companies) because "it's their product", others, particularly VDH officials, believed that communications from those within the biosolids profession may be biased or lack credibility. A few neighbors, both in Tulsa and the VA communities, mentioned the farmers because they have the knowledge and "can speak to the pros and cons". Interviewees suggested a number of other people who should be involved in informing the public, including government or city officials, the medical community, grassroots organizations and/or researchers.

When asked *how* the organizations who produce or use biosolids could best communicate about biosolids, interviewees cited a number of direct, local sources of information, including mailings, newsletters, brochures, email, public and/or "booths at home shows" and one-on-one meetings. Local media was also mentioned. Interviewees both in Tulsa and in VA, when asked, indicated that they would like a sign posted in a field before a biosolids application. While Tulsa interviewees suggested email, Internet or websites as a good way to communicate about biosolids, none of the neighbors interviewed in VA commented on having seen or read any websites specifically related to biosolids.

3.3.6 Comparing Research Results to Previous Research and Stakeholder Hypothesis

The research findings generally supported the prior research findings and the qualitative stakeholder hypotheses developed in Step 2 (Chapter 2.0). This study complements that prior research by generating empirical data into neighbors' thinking, adding rich context to those studies. The findings support Beecher et al.'s assessment of public acceptance issues (Figure 2-2 in Chapter 2.0), identifying a range of issues (e.g., odor and water quality) that concerned citizens in the local communications and stressing the importance of trust, equity and fairness and communications. This research suggests that uncertainty about emerging issues (which emerged in only a few of Beecher et al. case studies) will continue to emerge as an issue of concern.

The findings in this research also support the general research by risk perception researchers (e.g., Slovic, Fischhoff, and Lichtenstein, 1982) on the influences on the stakeholders' judgments of the acceptability of technology and other risks (See Table 2-1 in Chapter 2.0). Again, the research showed the importance of trust, clear benefits and individuals' sense of control and fairness on landowners' and neighbors' judgments. For example, landowners clearly weighed the benefits of biosolids over the risks and costs (e.g., odor is "short-lived" and "worth it"). Neighbors and VDH officials who were "in favor" of biosolids demonstrated a similar weighing of benefits or risks. Neighbors who were "not in favor" or "undecided", however, focused much more of the dreaded consequences, potential risks to children, involuntary exposure and unfair distribution.

Finally, the research findings generally supported the qualitative hypotheses developed by the researchers in Step 2 of the process (see Chapter 2.0) For example, supporting the hypothesis that n communities where biosolids land application has been successful, community members (which would include landowners and near neighbors) generally understood the basic process of treating wastewater and converting the residual sewage sludge to biosolids. Interviewees who were "in favor" of the local biosolids land application programs were familiar – and comfortable – with wastewater by-products being applied as biosolids on their local land.

Supporting the hypothesis that in communities where biosolids land application has been successful, risk managers have actively engaged community members and worked with them to build understanding of biosolids and the opportunity for beneficial use, the research showed that the Tulsa neighbors do not yet support the idea of the beneficial use of biosolids in their community – even though the program has successfully operated for over 25 years in their communities. The City of Tulsa, who has not had taken an active role in communicating with neighbors and other communities members, now has the opportunity to more proactively engage in dialogue with these community members.

One difference between this research and the prior biosolids perception research was interviewees' familiarity with the local land applications, particularly the landowners and neighbors from Tulsa and Fauquier County in Virginia, where the biosolids land application programs are well-established. Another key difference was the perceptions and judgment of biosolids land application by the Tulsa landowner interviewees, who were supportive of the land application. The research sample included landowners who have expressed interest to the City of Tulsa to participate in the program (e.g., on a 'waiting list'). These findings differed from Krogmann et al.'s (2001) study, which showed that a sample of New Jersey farmers were uncomfortable with biosolids land applications.

CHAPTER 4.0

DEVELOPING THE PRIMER FOR BIOSOLIDS PROFESSIONALS

4.1 Overview of the Adapted Strategic Risk Communication Process for Community Outreach and Dialogue on Biosolids Land Application

Using the key learning from the Research Challenge described in previous chapters, and the collaboration with the case study teams, the researchers customized the Strategic Risk Communications Process specifically for conducting outreach and dialogue with community stakeholders on biosolids land application. The resulting **Strategic Risk Communications Process for Outreach and Dialogue on Biosolids Land Application** was applied through the two case study applications (City of Tulsa, OK, and Virginia Biosolids Council) as summarized in Appendix B and Appendix D, respectively.

It became clear in the course of this work that biosolids managers need an actionable process that they can adapt to fit their own unique circumstances – recognizing wide variation in biosolids operations and communities across the country. Thus, the key considerations in adapting the process were that:

- The process can be tailored to fit the unique needs of a community and/or a specific issue or challenge biosolids professionals are addressing with key community stakeholders. These stakeholders may include landowners, people who live near land application sites and other community officials.
- The process can be adapted to address a specific project (e.g., a particularly sensitive land application site), or to provide an approach for an organization that conducts land application in many communities.
- The process and its steps can be scaled to provide maximum value with the available time and resources.
- The steps and guidance are clear and actionable.
- It supports continuous learning and continuous improvement. Because it is a systematic process, it can be applied repeatedly, building knowledge upon knowledge, and tracking how the issues, priorities, perceptions and communications needs of your stakeholders change over time.

The adapted **Strategic Risk Communications Process for Outreach and Dialogue on Biosolids Land Application** offers a set of six integrated steps and procedures that guide the implementation of critical tasks. See Figure 4-1. Step detail follows the figure. As shown in Figure 4-1, at any step in the process, the team may need to go back to an earlier step in the process, or (rarely), end the process before its completion.

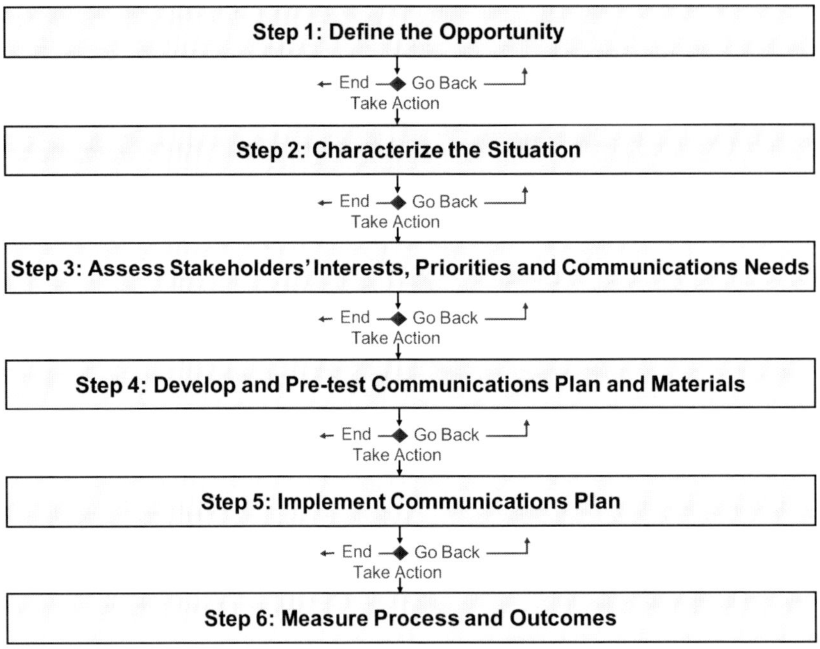

Figure 4-1. Strategic Risk Communications Process for Outreach and Dialogue on Biosolids Land Application.

- **Step 1: Define the Opportunity:** A Biosolids Communications Team, sponsored (and perhaps led by) the program manager, is formed. The team develops an Opportunity Statement – a clear statement of the current community outreach and dialogue opportunity for the biosolids program, including its goals, objectives and expected outcomes. The team also develops a plan for implementing the process, and (optional) guiding principles for community engagement and dialogue.
- **Step 2: Characterize the Situation:** The team conducts a preliminary technical assessment to understand the necessary scope and context of the biosolids land application program in communities, defining the various related technical, social, regulatory, environmental, economic, safety considerations. The team also conducts a stakeholder hypothesis to determine who the key stakeholders are, and the team's current understanding of those stakeholders' interests and priorities related to biosolids land application in their community.
- **Step 3: Assess Stakeholders' Interests, Priorities, and Communications Needs:** Using a variety of methods and tools – formal and informal research – the team gains an in-depth and first-hand understanding of key community stakeholders' values, priorities and interests; perceptions of biosolids and biosolids land application; judgment and decisions on the acceptability of biosolids land application in their community, and the influences on those judgments; and their communications needs, that is, what they want to know and how they want to learn and engage in dialogue about biosolids land application.

  This understanding is compared to what was learned in Step 2 (Technical Assessment and Stakeholder Assessment) and the team defines the gaps and alignments.

- **Step 4: Develop and Pre-test Communications Plan and Materials:** Using the insight generated in the previous steps, the team develops a communications plan for community outreach and dialogue that addresses what the team learned about people's priorities, perceptions and communications needs.
- **Step 5: Implement Communications Plan:** The team implements the Plan developed in Step 4. They monitor, adjust and refine the plan and materials as required, based on what they learn through their outreach efforts about stakeholders' interests and priorities.
- **Step 6: Measure Process and Outcomes:** The team measures the effectiveness of their outreach and dialogue efforts: to what degree do they achieve the opportunity defined in Step 1? They also assess the effectiveness of the process and make recommendations for future team efforts as part of their commitment to continuous learning and continuous improvement.

4.2 Development of the Primer for Biosolids Professionals

Drawing on the Health Canada Framework and Handbook, the researchers developed a *Primer for Biosolids Professionals: Conducting Effective Community Outreach and Dialogue on Biosolids Land Application* (referred to as the Primer in this report). The Primer was customized to meet the specific needs of professionals, which were discovered through the research and case studies undertaken for this Research Challenge. It provides the principles of risk communications and guidance on the tasks and tools to use in each step of the process. It includes concrete and actionable worksheets that biosolids professionals can use to complete the step. In many cases, the sample worksheets in the Primer are partially completed, drawing on the experience and learnings gained from the Research Challenge. Blank worksheets are provided in Appendix F.

One learning was that the process needed to be scalable to maximize its value with the time and resources available to professionals. Thus, the Primer includes guidance on how the process can be adapted in cases where resources are limited. The Primer also contains samples taken directly from the case study learnings.

4.3 Dissemination and Training

The Primer provides a detailed set of guidance, tools and examples that should guide a biosolids professional who is committed to working through it in detail. However, a key opportunity for WERF and its subscribers, along with members of other biosolids and wastewater management organizations, is to establish and conduct training for current and future biosolids professionals. The recommendation is to develop a comprehensive training package using the Primer as the training handbook. The purpose of the training would be to provide hands-on, in-depth training and coaching on how to operationalize the Strategic Risk Communications Process.

The Primer has been developed in a way so as to facilitate the development of training modules (e.g., through presentation slides). Typically this training would be of one day to a day-and-a-half in length. Participants would learn the essential principles, process and tools of Strategic Risk Communications, along with the rationale for their use. Session trainers would provide relevant examples and instruction to enable participants to have a well-rounded understanding and capability. Hands-on training with coaching on the application of the methods and tools to a current challenge faced by each of the participants would be the focus of most of the training. Each participant would leave the session having completed the first two steps of the process, with a plan for completing the rest in hand. Often one-on-one follow-up coaching is recommended as the participants work through their case.

Training on the application the Strategic Risk Communications Process could follow with more focused dialogue skills training for the professionals, and others on the team who will be conducting the outreach and dialogue. The dialogue training is typically a day-long session focused primarily on skills practice. The need for strong dialogue skills is described in the next section.

4.3.1 Building and Strengthening Dialogue Skills

Effective outreach and dialogue based on risk communications is not simply about the words used. Actions also send powerful messages. In most situations, especially those that can be characterized as having low trust and high concern, actions speak much louder than anything that is said. People often watch more than they listen and tend to draw more meaning from behavior than from what is spoken.

Therefore, skills training is critical. Formal risk communications training focusing first on planning, then developing and practicing dialogue skills, including active listening, is recommended. This training should be conducted by experienced practitioners who have the credentials to teach the science-based Strategic Risk Communications approach adopted by WERF, along with practical experience that is relevant to WERF subscribers. The purpose of such training is to enable the biosolids team members to design, conduct and evaluate effective outreach and dialogue with their key stakeholders.

Desired outcomes for such dialogue should include the following:

- Enable the team members to conduct dialogue with confidence.
- Enable team members to proactively engage stakeholders in dialogue about proposed biosolids land application and listen to and learn about their interests and priorities.
- Enable stakeholders to make well-informed decisions about the acceptability of biosolids and their application in the community.
- Provide stakeholders with multiple channels to share comments and suggestions along with their issues and priorities.

In general, training goals for each team member would include:

- Clear understanding of their respective roles.
- Demonstrated basic knowledge about biosolids and biosolids land application.
- Fundamental understanding of key principles of Strategic Risk Communications and the process and practices of conducing effective consultation through outreach and dialogue.
- Demonstrated ability in dialogue skills.
- Readiness to conduct outreach and dialogue following the training.

While formal training is ideal, it may not be an option in some cases. Informal coaching by the team members and perhaps supported by others in their organization can also be beneficial in building team members' skills and confidence.

CHAPTER 5.0

OUTREACH, RECOMMENDATIONS, AND FUTURE RESEARCH NEEDS

5.1 Research Challenge Outreach

The researchers have had several opportunities over the course of this work to conduct outreach with and engage biosolids professionals. they have benefitted from their engagement with the Research Challenge Biosolids Risk Assessment Team collaborators, the Issue Area Team, and the Science and Regulatory Advisory Panel (convened by the Research Challenge Biosolids Pathogen Risk Assessment Team collaborators). This research was also presented at the 2008 and 2010 WERF Research Forum Meetings.

5.1.1 Workshop on Communications on Biosolids Land Application

Of particular relevance was a half-day workshop on "Assessing and Communicating Risks from Pathogens in Land Applied Biosolids" which was jointly organized by the Research Team and our collaborating Biosolids Pathogens Risk Assessment Team,conducted as part of the Water Environment Federation's (WEF) 2010 Residuals and Biosolids Conference in Savannah, Georgia. This workshop provided a forum for the discussion of appropriate biosolids management practices. See Appendix G for the agenda and workshop slides.

The 2010 WEF workshop included 34 participants from the wastewater and biosolids profession, along with academics and researchers. The workshop was participatory in nature. The workshop began with a plenary presentation overview of the stakeholder engagement and communications on biosolids land application, followed by an overview of pathogen risk assessment andthe SMART Biosolids tool developed by the Biosolids Pathogen Risk Assessment Team. This was followed by three presentations describing the challenges and opportunities facing biosolids professionals in three communities or regions (Tulsa, OK, Washington, D.C., and Pennsylvania).

After the presentations, participants were led in discussion in breakout groups facilitated by the workshop presentators.The focus of the breakout discussions was opportunities and challenges related to stakeholder engagement and risk communications of biosolids land application. Key group and plenary discussion topics included: developing and maintaining consistent messages; building stakeholder trust; gaining organizational and regulatory partnerships and buy-in; pre-planning a communications strategy; maintaining respectful partnerships with media; creating emergency response and crisis communications plans; and engaging stakeholders through outreach, dialogue, site tours, educational outreach and relevant print materials. Participants were encouraged incorporate what they had learned in the presentations and how that might apply in their own progarms' circumanstances.The breakout group discusions were reported back to the large group session for a large group discussion on breakout group reports.

The success and positive feedback from the participants at the 2010 WEF workshop leveraged a second opportunity to conduct a similar workshop scheduled for September 14, 2010 in Clemmons, North Carolina under the sponsorship of the North Carolina Water Environment

Association. About 20 municipal utility staff, land application contractors and consultants attended, primarily from North and South Carolina. Finally, the workshop and the plenary presentation on Strategic Risk Communications were highlighted on the Mid-Atlantic Biosolids Association's website.

The success and positive evaluations of the workshops support the idea that biosolids professionals across the U.S. will see the value in more systematically and strategically conducting outreach and dialogue within their communities. It also suggests that biosolids professionals will be open to learning more about the Strategic Risk Communications Process and working with its methods, tools and guidance. The workshop also provides the foundations for the development of a training program to help biosolids professionals customize and adopt the process in their own programs.

5.2 Applying the Process to Communicate About Pathogen Safety

Although the focus of this Primer was on developing community outreach and dialogue on biosolids land application, the process could also be used to address a number of more specific challenges and opportunities facing biosolids professionals. This includes effectively communicating with stakeholders about pathogen safety. With the advances of risk assessment in pathogen risk of biosolids (i.e., the development of microbial risk assessment tool by the WERF Research Challenge Biosolids Pathogens Risk Assessment Team), there is a significant opportunity for WERF, its subscribers and partner organizations, to coordinate efforts in developing plans, guidance and supporting communications materials to facilitate effective dialogue on the pathogen safety of biosolids land application.

The Strategic Risk Communications Process can be applied to help achieve this opportunity. The application of the process through the two Research Challenge case studies can serve as the foundation for providing a broad understanding of a sample of stakeholders' priorities on biosolids safety. A key learning from these cases is that biosolids safety is a top priority for all stakeholders, particularly neighbors to land application sites and, in the case of Virginia, health officials (VDH). How these stakeholders conceptualized safety was much broader than pathogen risk. neighbors, for example, referred to "contaminants" and "unknowns". They were concerned about metals, hazardous waste and pharmaceuticals. VDH officials said that they were "comfortable" with pathogen risk and were more concerned with long-term health impacts of land application.

Communications to address the specific needs of these and other stakeholder must provide information that is personally relevant. And it must start by addressing where people are today in their thinking on safety, pathogens and risk. For example, while "experts", in this case, people in the profession, typically want to emphasize the facts (e.g., the numbers, the data, the regulations) in their communication, laypeople typically want to speak first of their values – what is important to them – and have those acknowledged and addressed by experts. Values important to these and other stakeholders typically include the health and well-being of their families, fairness, and due process – that they have been invited to participate in a meaningful way in decisions that affect them. While it is important to share the facts and data, it must be done in a way that is relevant and understandable to laypeople, while addressing their values, interests and priorities. The data will not carry the day.

With that in mind, how can the process be applied to develop effective communications on pathogen risk? The first task as part of Step 1: Define the Opportunity would be to form a Biosolids Communications Team. This team could be sponsored by an organizations such as

WERF, the Water Environment Federation and/or the National Biosolids Partnership. The team would also need to include a key expert on pathogen risk and risk assessment and some key program managers who are in charge of managing those risks. It might be beneficial to include other professionals, experts, regulators, etc. on the team.

The next part of Step 1 would be to develop the Opportunity Statement. The team would need to consider the following:

- Who do we need to engage in a dialogue about pathogen safety and what would be the primary purpose of this effort? For community members, the purpose may be to build positive judgment and confidence in the safety of biosolids land application in their community. Another desired outcome, for example, might be to use this outreach and dialogue on biosolids as a critical step in helping to shape policy at the local and/or state or federal level. It is important that the opportunity be as focused as possible, and it must have clear outcomes that are measurable and achievable.
- How will the opportunity be achieved? The key to designing and conducting effective community dialogue is to start by discovering community stakeholders' interests, priorities, perceptions and the desired communications – what they want and need to know to make a well-informed decision about pathogen safety – then to effectively address these through communications tailored to their needs. This insight can be revealed through formal or informal research designed specifically for this purpose. Using the key learnings from this research, the team can develop a focused communications plan centered on engaging key community stakeholders in dialogue. They can also develop appropriate and relevant support materials that are aligned with the interests and needs of community stakeholders.
- What is the scope of such an initiative? The Research Challenge focused on the development of community-level outreach and dialogue. The team may determine that the opportunity to design and conduct effective communications on pathogens should have more of a national focus, that is, it should focus on landowners, neighbors and other key community stakeholders in communities across the nation where biosolids operations are underway. The determination of the scope should be confirmed after the team completes Step 2.

The next step, Step 2, would be to Characterize the Situation, including:

- A Technical Analysis – what is known about pathogen risk and risk management? It will be important to integrate what is known about pathogen risk. A summary of expert knowledge – what is known about pathogen safety, what is still unknown and how pathogen risks are managed – should be compiled in this step. The biosolids microbial risk assessment tool developed by the WERF Biosolids Pathogen Risk Assessment Team would be particularly useful to the team. The Expert Model of Influences on Integrated Risk Assessment and Risk Communications (RA/RC) System for Biosolids Land application, which was developed as part of the WERF Research Challenge (Appendix B) could also be a very important resource.
- A Preliminary Stakeholder Assessment – who are the key stakeholders, and what does the team currently understand about their interests, priorities and communications needs? In this case, key stakeholders would include the farmers and

near neighbors to an application site because they are the ones who are closest to the risks. Other key stakeholders might include employees and Department of Health officials.

In Step 3, the team would design and conduct research with a sample of key stakeholders in order to understand, in depth, their priorities, interests and communications needs. Because the topic of pathogen risk is complex, it is recommended that formal research methods be applied. The research design presented in this WERF Research Challenge could serve as a foundation. It could be tailored to include more specific questions about: what people think about pathogen risk in general, and specific to biosolids; how they compare pathogen risks in biosolids with other risks and what tradeoffs they are willing to take; what they consider to be "acceptable" risk of pathogens and why; what they expect from decision makers regarding the management of pathogen safety; their current confidence in pathogen safety and the factors that would increase their confidence in the safety of biosolids with regard to pathogens; and what additional information they want to have about pathogens and how they would like to receive that information.

Although not conducted as part of this Research Challenge, qualitative mental models research is often followed with quantitative studies, building on what was learned in the qualitative research. The goal of such research is to better understand the prevalence of thinking in a population of stakeholders, for example, neighbors in communities around the nation. This might be an appropriate extension of the research contemplated here.

With the insights gained from Steps 2 and 3 the team could use the guidance and tools in Steps 4-6 in the Primer to develop, pre-test and implement communications plans and materials, and then evaluate the process outcomes. (The technical FAQ in Appendix C-3, drafted as part of the Tulsa case study, provides an additional foundation for research design and communications development.)

5.3 Specific Guidance on Using Risk Comparisons to Communicate About Biosolids Benefits and Risks

A key question that was repeatedly raised in the course of this work was the appropriateness of the use of risk comparisons (e.g., comparing the risk of biosolids with other risks, such as produce contamination, recreational waters and drinking water).

Strategic Risk Communications help people make informed decisions by providing personally relevant information specific to the decisions they face. Done well, the can help make unfamiliar benefits and risk information more comprehensible, accessible and relevant to people. Comparisons are often recommended when statistical data alone will not help people judge the acceptability of, or make choices about, benefit and risk. They are particularly useful to highlight various aspects, dimensions, or components of a benefit or risk, and/or to help people compare unfamiliar risk to more personally relevant risks (Roth et al., 1990).

While risk comparisons may be an effective way of putting a risk into context, they also come with a unique set of challenges. If not carefully applied, risk comparisons can hinder, or even reverse, the intended effects of the risk communications effort. As noted by a Carnegie Mellon risk communications research team, "a poor risk communication can often cause more public (and economic) damage than the risks that it attempts to describe (Morgan et al., 1992, p. 2055)". These unintended consequences may be exacerbated by a lack of trust or credibility in the communicator (Johnson and Chess, 2003).

Consequently, any use of risk comparison calls for careful consideration of the context of risk judgments and resulting decisions. Effective risk comparisons require a clear understanding of communication goals and objectives, as well as an awareness of the needs of the recipients and an understanding of the way that recipients will process the message content (Santos and McCallum, 1997). Table 5-1 provides some general guidelines on the use of risk comparisons, based on research and expert guidance (Fischhoff 1995; Johnson and Chess, 2003; Roth et al., 1990; Thorne et al., 2006), that are relevant to communications on complex topics addressed in biosolids management.

Table 5-1. Risk Comparison Do's and Don'ts.

Risk Comparison Do's:

- *Always* pre-test risk communications messages, especially risk comparisons.
- Carefully consider the goal of the risk comparison – how is this message intended to inform people's judgments or decisions related to biosolids management.
- Use risk comparisons as a means to help people *understand* the risk (e.g., size, outcomes, etc.) or to make the information more personally relevant. Risk comparisons should not be used as a means to persuade people to accept or not accept the particular risk.
- Ensure that messages are truthful, accurate, and based on solid evidence. Risk comparisons have the added burden of requiring accuracy of both the focal and the compared risks.
- To the extent possible, compare risks that share common factors that influence people's acceptability. Use comparisons of unrelated risks or verbal analogies *only* to convey the magnitude of risk.
- Beware of using risk comparisons if the risk is highly uncertain or scientific understanding of the risk is limited **and** the comparison would neglect or de-emphasize this uncertainty.

Risk Comparisons Don'ts:

- Don't impose judgments of risk acceptability on the recipients (e.g., "if you accept Risk A, then you ought to accept Risk B (Roth et al., 1990))."
- Don't complicate messages by including risk comparisons.
- Don't risk not pre-testing messages.

5.4 Key Observations from the Research Challenge

The Research Challenge demonstrated that long-term the viability of biosolids land application depends on continuously earning stakeholder confidence, trust and support. Over the course of this Research Challenge, it became clear that a fundamental opportunity for biosolids professionals – the people and organizations responsible for the safe production and appropriate use of biosolids – is to first define and then demonstrate a strong commitment to their people, their products and services, and the people in the communities in which they operate. How? Through their management orientation.

Management orientation refers to the integrated values, beliefs and attitudes that guide the judgments, decision making and actions of members of the people in the broad range of organizations that comprise what may be called the biosolids industry. Every day the biosolids professionals who comprise this industry demonstrate their management orientation through their communications of all kinds – what they say (their words) and what they do (their actions).

As the work on the Research Challenge is concluded, the researchers offer some key considerations on elements of a management orientation that can serve biosolids industry leaders and other biosolids professionals as they work to improve their communications, outreach and dialogue with their stakeholders. These considerations are drawn from their years of experience, including interactions with a wide variety of biosolids professionals, and the specific research results of the WERF Challenge.

Specific recommendations for designing and conducting effective outreach and dialogue, based on the Strategic Risk Communications process, can be found in the *Primer for Biosolids Professionals: Conducting Effective Outreach and Dialogue on Biosolids Land Application* (referred to as the Primer), which was a key deliverable of this Research Challenge and is available through the Water Environment Research Foundation.

While aligned with the practical recommendations in the Primer, the considerations below focus on the higher-value opportunities for biosolids professionals and their organizations. These include the following:

- Striving to systematically earn trust and, ultimately, societal support by actively demonstrating: a) your commitment to your people, products and communities; b) good performance; and c) your efforts to communicate effectively.
- Reframing biosolids as a product, not a waste. By reframing, we mean redefining how you think about and communicate about biosolids – that is, biosolids are a valued product that provides a range of benefits to farmers, communities and society versus biosolids are a waste that society must deal with.
- Ensuring that good performance precedes your communications. As many industries and companies have learned the hard way, you can't communicate your way out of bad performance. Addressing performance issues – in operations, product quality, or the work of your people – must come before any communications about your performance as a company as demonstration of your efforts to continue to earn community support and your ongoing license to operate.
- Conducting outreach and dialogue with your community stakeholders in a way that is supported by principles and guided by insight.
- Being proactive and leading the dialogue on biosolids land application.

5.4.1 Trust Trumps All and Must be Earned Continuously

The research undertaken for the Research Challenge underscores that the key influence on stakeholders' judgment of the acceptability of biosolids land application is the degree to which they judge the people and organizations producing and applying biosolids, and their communications, to be trustworthy and competent. This means that the demonstrated competence of the biosolids program alone cannot earn stakeholders' confidence if the people who work for the program are not believed to be considerate of stakeholders' values, interests, and priorities and, therefore, worthy of their trust.

Stakeholders' judgments of the trustworthiness and competence of the people and organizations managing biosolids must be *earned* through good performance and effective communications. Trust is fragile. It is hard to earn, and it can be easily lost through careless communications or performance. Earning trust is a never-ending process. It requires a commitment from leaders at all levels of an organization and throughout the industry.

5.4.2 Biosolids: A Product, Not a Waste

A key observation from the Research Challenge is the opportunity to change the framing – in management orientation and in communications – of biosolids as a *product*, not a waste. The difference between the two perspectives is profound.

Products are associated with *value*, typically intended to provide benefits. Wastes are not. With any product, there is an exchange of value between the producer and the receiver, or buyer. One consideration that arose in the course of this research was framing biosolids as a beneficial product. This was a key discussion topic in the two case studies conducted as part of the Research Challenge and significantly contributed to the development of the case study teams' community engagement *Opportunity Statements.*

For both the biosolids professionals and their companies and organizations, a positive attitude about biosolids as a valuable product along with progressive beliefs and positive actions can produce real outcomes in the form of successful strategies and communications. Being proud and positive about the products you make and the people who enable its beneficial use will have significant positive results.

The re-framing of biosolids as a product requires alignment and commitment at all levels of its production and application, as well as through all communications.Those who make products and guide their appropriate use are typically expected by others – stakeholders and customers alike – to be responsible for them, often throughout their lifecycle. As such, they are:

- Held to high standards by all stakeholders for product quality, performance, and product safety.
- Held to high standards for communications, including knowledge, credibility, openness, transparency, and accessibility. They are expected to listen to and learn from stakeholders, and provide clear and balanced communications about their products and operations.
- Expected to demonstrate that they have heard and *care* about the interests and priorities of all of those affected in some way by their products and operations.

The research demonstrated that biosolids programs' key stakeholders hold the same high expectations of the biosolids industry and the people who comprise it.

As part of this management orientation, a specific consideration for the biosolids professionals is to reconsider providing "free" product to landowners, which sends a message that biosolids have no appreciable value. This devalues the product and sends a confusing message to landowners, community stakeholders and others as the producers of biosolids communicate the range of benefits of biosolids for land application. In many communities, like the City of Tulsa, for example, demand for biosolids outstrips supply. If biosolids were universally viewed as a product, why could this demand not increase across the United States?

5.4.3 Performance Precedes Communications

There is an axiom in the practice of communication that can serve as a strong guiding principle for management by biosolids managers; namely, performance must precede or parallel communications. Performance means how the biosolids professionals in the industry run their businesses and treat their stakeholders. Actions communicate. People take meaning from what they see and what they don't see. What people observe can have as much or more of an effect on their thinking as "messages" they receive through communications. Stakeholders can be harsh judges of other people's professionalism and motivations if they believe they have been mistreated, or treated unfairly.

All aspects of biosolids operations can be a form of communication. In practical terms, this means that poor housekeeping at an operating facility or storage yards with sloppily stored and rusty equipment (performance) can be interpreted by people as "evidence" or "signals" of incompetence, lax management and/or unsafe operations. By the same token, clean and well maintained trucks operated well by people who appear to be professional can send a strong message of competence and commitment to appropriate standards. Moreover, if properly communicated, the value of your product can speak for itself. Demonstrated benefits of biosolids land application – flourishing fields and reclaimed land – not only sell the product, but are a positive reflection of the product and the people who produce it.

Biosolids professionals should expect people to question their communications and motives for communicating. Consequently, it is unlikely that communications, however well intended, will perform as intended if the performance of your product, people or operations is not up to standard. Biosolids professionals must understand that there is no way to successfully "communicate" or "PR" your way out of an issue (like odor or a management failure), especially one related to inadequate performance.

5.4.4 Leverage the Advantage of Being Proactive

Open, proactive communications can result in important advantages. By virtue of being proactive, biosolids professionals are better able to establish the agenda of topics for dialogue with stakeholders. And they can more ably set and maintain the pace of communication.

The opposite is also true. Avoiding communication by "flying under the radar" of stakeholder interests, or providing incomplete or inaccurate information to stakeholders, are communications strategies that can backfire with potentially severe results in terms of loss of biosolids professionals' credibility and damage to the viability of their programs and possibly those of others in the industry.

A specific consideration, in light of the case study applications, is to develop strategies to communicate about biosolids land application through onsite signage. The case study teams identified signage as an opportunity to be more open and transparent about their operations, and to demonstrate their pride in their products and services. They believed (suggested by the research results, although further testing is required) that signage would have merit in their communities. In addition, a proactive step in developing and using signage could be an important voluntary initiative that could either a) offset the need for regulated signage or b) be accepted by lawmakers as appropriate signage.

5.4.5 Communications Should be Guided by Insight

The purpose of the Strategic Risk Communications Process is to guide – through insight – the industry's outreach and dialogue with its key community stakeholders. The research demonstrated that developing this insight is especially critical because the communication environment for biosolids is complicated. Biosolids professionals might easily pre-suppose or guess about the issues for stakeholders and, therefore, misjudge, however inadvertently, what is important to stakeholders and what to emphasize in communications. For example, in this research, the safe use of biosolids was a focal point for stakeholders. The perceived that the unpleasant characteristics of biosolids, and the uncontrollability of these after application (as in a potential to contaminate runoff waters), were also important to community stakeholders. Pathogens, their presence in biosolids and risks presented by them, were not, or not always, identified and key topics of interest to stakeholders.

Understanding specific community stakeholders' needs, interests and priorities is critical because there is not one message, or set of messages, on biosolids land application that will meet the needs of all stakeholders, all communities, or all situations. This Research Challenge demonstrated that interests and priorities of stakeholders varied widely within each community involved in the case study applications.

5.4.6 Lead the Dialogue, Taking the Higher Ground

A final consideration for the biosolids professionals, and the biosolids industry more generally, is to lead the dialogue by taking the higher ground. This means opening the conversation about the values that are shared by stakeholders (environmental quality, for example), and the higher order benefits regarding biosolids that are consistent with these values (community health and well-being and environmental stewardship, for example). This is especially important in times when the dialogue on biosolids becomes adversarial. By so doing, biosolids professionals can demonstrate that they are aligned with the values and priorities of the people in the communities where they have operations.

Taking the high ground in communication means demonstrating that you have listened to – and respect – your stakeholders and their interests. It means that you openly acknowledge a balance of perceived benefits with perceived downsides or challenges associated with biosolids, and thus avoid being perceived to be defensive or argumentative. By taking the high ground, biosolids professionals can demonstrate this respect and be even-handed in how they communicate about topics related to biosolids.

5.5 Future Directions

With the development of a systematic process, actionable tools, step-by-step guidance, and in-depth demonstrations resulting from this Research Challenge, biosolids professionals should be much better prepared to design and conduct effective communication with their key stakeholders – the people whose decision making affects the success of their biosolids programs.

Helping biosolids professionals adopt the process as a leading practice is the next crucial step. Training, such as that described in Chapter 4.0, will be key to building competence and confidence in this approach. To maximize the benefit of this approach, biosolids professionals will need to gain the necessary support, buy-in and resources from their key partners to fully implement the process and realize its benefits. For municipal utilities managers, as an example, this may require buy-in from municipal governments or state regulators.

Continued applications of the process will generate increasing value and position the process as a leading management practice for biosolids professionals. The process can also be applied and make significant contributions to water utility professionals' efforts on wastewater, stormwater and other environmental management programs.

Another opportunity is to conduct research with biosolids professionals using this approach, to assess their successes, gather best practices and key learnings and also identify what they want and need in order to be more able and confident in their stakeholder outreach and dialogue efforts. This research could guide the development of specific training, guidance, coaching, workshops and/or lessons learned summaries that would support biosolids professionals, and the industry as a whole, in becoming more capable and competent in communications. In this way, the initial investment by WERF in this Research Challenge will be leveraged to support capacity building and continuous improvement in the industry.

A second opportunity to leverage this Research Challenge is to continue – through application of the process – to generate empirical insight into stakeholders' priorities, perceptions and communications needs on biosolids management. These research efforts should focus on key stakeholder, such as neighbors, landowners, health officials, etc. Using the case study applications as a base, similar research can be conducted in other communities across the U.S., Canada, and other countries, building a more complete picture of stakeholders (including the public's) current thinking on biosolids. The expert models/mental models research method is specifically designed to enable the integration of research across communities.

Finally, the work conducted as part of this Research Challenge, coupled with that of the development of the pathogen risk assessment tool by the Biosolids Pathogen Risk Assessment Team, can also support a continued effort by WERF, its partners, and its subscribers, to advance a integrated risk management and risk communications approach to biosolids land application (Beecher et al., 2005). An integrated risk management and risk communications framework will become even more valuable as WERF and its partners continue to contribute to the development of other management tools and best practices, such as the Biosolids Odor Reduction Roadmap tool (BORR–03CTS9c) and incident response protocol (08HHE5PP).

APPENDIX A

EXPERT MODEL NARRATIVE

A.1 Introduction

The goal of this WERF Research Challenge (SRSK2R08) was to develop a process, methods, tools, and materials to enable biosolids managers to conduct effective outreach and dialogue with key stakeholders in the communities where they operate, on both biosolids land application and its beneficial use, and on specific challenges such as potential health risks of pathogens in biosolids. Through this Research Challenge, the science-based **Strategic Risk Communications Process for Outreach and Dialogue on Biosolids** was developed, which biosolids professionals can adapt as they design and conduct stakeholder outreach and dialogue in their local communities. The process, originally developed for Health Canada and the Public Health Agency of Canada, was modified to fit the unique (and often unmet) communications needs of biosolids professionals.

Effective communications strategies, including outreach and dialogue on biosolids, must reflect the best available (expert) understanding of the complex topic or management issue at hand. Equally important, they must be based on an in-depth understanding of stakeholder interests, priorities, and information needs related to those topics. Thus, successful adoption of the Strategic Risk Communications Process should include sound and purposeful research with both experts and stakeholders.

The Expert Model/Mental Models research method (Morgan et al., 2002) is ideally suited to generate needed insight in both the expert and stakeholder domains. The method was designed by risk analysis and risk communications experts as a means to discover and characterize people's thinking and information needs about a complex issues in risk management and inform the development of focused communications plans and materials that address those priorities and needs. The method was used by the researchers to provide the foundational knowledge to adapt the process and apply it in the two case study demonstrations. (See Chapter 1.0 for a more detailed description of the approach).

The first step of this approach is the development of an expert model:

Definition: *An expert model is a summary of relevant knowledge about a topic, typically illustrated in the form of an influence diagram. As a formal integration of knowledge, the expert model summarizes the knowledge needed to make judgments about the topic and issues related to it. It is prepared in a way that will facilitate risk analysis, risk communications and, ultimately, risk management.*[2]

This Expert Model Narrative summarizes the Expert Model and its development. Section A.1 presents the background on the methodology, and the steps taken to develop the Model. Sections A.2 and A.3 then present and describe the Expert Model. Section A.2 introduces a high-

---

[2] *From* *Strategic Risk Communications Framework and Handbook*. *Prepared for Health Canada and the Public Health Agency of Canada by Decision Partners, November 2006/Version 1/Volume 1/ (www.riskcommunications.gc.ca).*

level or Simple Model, while Section A.3 presents more detailed models, called Detailed Expert Sub-models.

A.1.1 Background on Expert Models

The objective of Strategic Risk Communications is to enable decision makers and stakeholders to make well-informed decisions that lead to responsible and ethical risk management. If the process of engaging stakeholders throughout this process is to be effective, it must start with shared understanding, at the expert level, of the context or system in which decisions about the acceptability of biosolids land application are made. Such understanding involves many complex issues, which must be systematically identified and addressed. These include:

- The context, or *system*, that characterizes biosolids land application. This includes the decision-making processes that influence how biosolids are produced and land applied, the benefits and risks of an application project, and other factors that influence those decisions.
- How stakeholders perceive and assess the benefits and risks associated with a land application project, and how they make tradeoffs between them.
- Stakeholders' informational needs related to potential biosolids applications, specifically what stakeholders want and need to know about a proposed land application project in order to make a well-informed decision about the project's acceptability.

One of the most effective ways to capture this understanding is by developing an *expert model*. As a formal integration of expert knowledge of a system, expert models are essential management tools. They enable identification of critical issues in risk assessment and risk management; gaps in current policies and procedures and opportunities to improve stakeholder engagement; and can be readily revised and refined over time to capture the most current understanding of key variables in the system it addresses. The development of expert models, typically through workshops with experts from a range of disciplines, can be an effective way to build shared understanding of the system being addressed.

An expert model also provides the analytical framework from which researchers can design, conduct, analyze, and report on stakeholders' *mental models*. Mental models are often described as the complex web of deeply held beliefs that guide people's judgments and decision making (Johnson-Laird, 1983, Morgan et al., 2002). Mental models research is used to discover stakeholders' values, perceptions, and informational needs – that is, what they want and need to know in order to make a well-informed decision on the topic at hand. Through mental models research, researchers can identify the gaps in stakeholders' understanding of the risks, benefits, and tradeoffs associated with a topic and, at the same time, identify gaps in experts' understanding of other influences on stakeholders' decision making. Thus, the research can inform the design of strategies and communications to effectively address stakeholders' values, needs, interests, and priorities about the topic.

A.1.2 Purpose of the WERF Research Challenge Expert Model

The expert model developed for this project is entitled **Expert Model of Influences on Integrated Risk Assessment and Risk Communications (RA/RC) System for Biosolids Land application.** The focus of this Expert Model is on assessing and communicating the risks of pathogen exposure from biosolids land application. However, it is clear that pathogen risk is one in a set of complex factors and cannot be managed – or communicated – in isolation. The Expert Model illustrates these factors.

The purpose of the Expert Model of Influences on Integrated Risk Assessment and Risk Communications (RA/RC) System for Biosolids Land application is to identify the range of:

- Stakeholders who can affect or be affected by biosolids land application.
- Decision-making processes for producing, storing, transporting, and applying biosolids, and the potential influences on that process, such as regulations, site conditions, and other factors.
- Factors that influence the risk of pathogens in biosolids land application, including the likelihood of environmental, animal, and human exposure to pathogens.
- Potential impacts from biosolids land application, including health, environment, economic, and political impacts.
- Potential influences on the effectiveness of risk assessment and risk communications processes, methods, and tools.
- Potential influences on key stakeholders' judgments of the acceptability of biosolids land application.

Note: This Expert Model was developed for the WERF Challenge as a generic model for biosolids land application. It is intended to be adapted by WERF subscribers to the specific challenges they face. The Expert Model has been adapted, for example, as part of the Tulsa Case Study.

A.1.3 Expert Model Scope and Definitions

The following terms are used in the Expert Model:

- Appropriate Use of Biosolids: Stakeholder and societal judgment of the degree to which biosolids are produced and land applied in a way that maximizes the benefits, while ensuring minimal risk, of the land application to people and the environment.
- Biosolids is defined as the primarily organic solid product yielded by municipal wastewater treatment that has been processed to meet the land application standards in the Federal Part 503 Rule.
- Quality: The criteria stakeholders use to evaluate a process or outcome, in this case, the acceptability of biosolids land application.
- Risk is the chance of something happening that will have an impact upon people or their environment. It is measured in terms of the potential consequences and likelihood or chance of those consequences occurring.
- Stakeholder is defined as any individual, group or organization that may affect, be affected by, or perceive itself to be affected by decisions related to biosolids issues, projects, applications, problems or policies.
- Strategic Risk Communications is defined as the purposeful process of skillful interaction supported by appropriate information to enable well-informed decision making and action on risks. It is an integral part of the risk management process.

A.1.4 Developing the Expert Model

Each expert model is unique to the situation at hand. The process of developing an expert model is an iterative one, as experts review draft models and make recommended improvements. The **Expert Model** was developed by Decision Partners in consultation the Research Team, WERF staff and the WERF's Biosolids Issue Area Team (IAT). The following tasks were completed in the development of the Expert Model:

Conducted a secondary review: Decision Partners started by reviewing current scholarly articles and regulatory and guidance documents related to biosolids treatment and land application, biosolids pathogen risk and risk management, community and public perceptions of biosolids and biosolids risk management, and general research on risk communications. The review included an Internet search on topics related to biosolids land application and pathogen risk. Decision Partners also consulted with individuals on the Biosolids Pathogens Risk Assessment Team to gain broad expert insight on biosolids land application and pathogen risk assessment.

Drafted a Simple Expert Model: From the insight generated through the secondary review, Decision Partners drafted the Simple Expert Model, which presents a high-level diagram of the "system" of biosolids land application. This model is described in Section A.2.

Facilitated expert elicitation: In August 2008, WERF hosted a joint meeting of the Research Team (including Decision Partners and the Biosolids Pathogens Risk Assessment Team) and the Research Challenge IAT. As part of this meeting, Decision Partners reviewed the Simple Expert Model and facilitated a discussion to gain insight needed to develop the Detailed Expert Model. Discussion topics included the influences on the need and opportunity for more effective biosolids land application risk assessment and risk communications processes; the range of stakeholders who may affect or be affected by biosolids land application; the range of processes for developing, storing, transporting, and applying biosolids; biosolids pathogen risks; and potential impacts of biosolids land application. Decision Partners also facilitated a discussion on Research Team and IAT members' hypotheses about the factors that influence stakeholders' judgment on the acceptability of biosolids land application.

Drafted a Detailed Expert Model: From the input generated through the secondary review and expert elicitation, Decision Partners refined the Simple Expert Model and drafted the Detailed Expert Model. The Detailed Expert Model consists of seven sub-models that are described in Section A.3.

Reviewed Expert Model with the Biosolids Pathogens Risk Assessment Team, IAT and WERF subscribers: In December 2008, Decision Partners presented the Simple Expert Model and Detailed Expert Model at the WERF Research Forum. WERF subscribers had an opportunity to comment. Decision Partners also presented the Expert Model to the IAT later that month. Based on the feedback received, the models were refined.

Reviewed Expert Model with Tulsa Biosolids Team: In May 2008, Decision Partners facilitated a workshop with staff of the City of Tulsa Public Works Department, the host utility for the first Research Challenge strategic risk communications case study. The models were reviewed and discussed, along with recommendations for refinements. These changes have been incorporated in to the models presented in this narrative.

A.1.5 Organization of the Expert Model Narrative

This Expert Model Narrative is designed to provide information to help guide the reader through the Expert Model by providing supplemental information associated with each of the Models. It is divided into the following sections:

- Section A.2: Simple Expert Model of Influences on Integrated Risk Assessment and Risk Communications (RA/RC) System for Appropriate Biosolids Land Application. This section presents the highest-level version of the Model outlining the key "nodes" or variables in the system and relationships among them. These variables include the range of influences from the stakeholders in the system, to biosolids production, biosolids land application, pathogen exposure, and impacts from biosolids land application.
- Section A.3: Detailed Expert Sub-Model of Influences on Integrated Risk Assessment and Risk Communications (RA/RC) System for Appropriate Biosolids Land Application depicts additional variables that influence the macro nodes. Because the Model is complex, it is divided into seven separate sub-models that focus on particular variables of the Simple Expert Model:
    - Detailed Expert Sub-Model A: Stakeholder Map of Biosolids Land Application
    - Detailed Expert Sub-Model B: Quality of Biosolids Production
    - Detailed Expert Sub-Model C: Appropriateness of Biosolids Land Application
    - Detailed Expert Sub-Model D: Influences on Degree of Pathogen Exposure
    - Detailed Expert Sub-Model E: Impacts of Biosolids Land Application
    - Detailed Expert Sub-Model F: Influences on Social Judgment of Biosolids Land Application

A.2 Simple Expert Model of Influences on an Integrated Risk Assessment and Risk Communications (RA/RC) System for Biosolids Land Application

The Simple Expert Model in Figure A-1 depicts the system of integrated risk assessment and risk communications for appropriate biosolids land application. The Model also illustrates how a Risk Assessment/Risk Communications (RA/RC) Process, with its supporting methods and tools, influences the Quality of Biosolids Production, the Degree of Pathogen Exposure, Social Judgment, Desired Outcomes, the Quality of Biosolids Professionals' Adoption of a RA/RC Process, Methods and Tools, and the Appropriateness of Biosolids Land Application.

A.2.1 How to Read the Simple Expert Model

The Expert Model is depicted as an *influence diagram*; that is, a directed graph in which arrows, or "influences" link related "nodes" – the variables in the system. An arrow between two nodes indicates that the node at the tail of the arrow exerts a certain "influence" on the node at the arrow's head.

The nodes in the Simple Expert Model are *macro nodes* that is, core variables in the system of biosolids land application. The bulleted lists within the nodes provide detail on additional types of variables included within the macro nodes. For example, the node *Quality of Biosolids Production* encompasses several processes, like treatment and secondary production.

Nodes with a dashed border indicate that the node is presented and described in further detail as a Detailed Sub-model (See Section A.3).

Note that this Simple Expert Model is a systems model – not a process flow diagram or technical risk model. It identifies the key factors that must be considered in an effective RA/RC system for biosolids land application (e.g., pathogen factors, such as virulence and persistence). However, it does not model technical relationships among these factors.

A.2.2 Overview of Simple Expert Model

The Simple Expert Model begins in the upper left-hand corner with the *System Drivers* and ends in the lower right-hand corner with *Desired Outcomes* of the proposed integrated risk assessment and risk communications system for appropriate biosolids land use. *Drivers* represent the higher-level influences motivating the need or opportunity for biosolids land application and, more specifically, ensuring appropriate risk assessment and, ultimately, risk management of pathogens. *Desired Outcomes* represent the ultimate objectives of the RA/RC system – to improve decision making on the sustainable use of biosolids in land application.

The other nodes link the *System Drivers* to the *Desired Outcomes* through a complex set of relationships. These nodes depict the various stakeholders in the system, the processes for producing and land applying biosolids, the RA/RC process, methods and tools to guide those processes, factors that influence the risks of pathogen exposure and the potential impacts of biosolids land application (including those driven by the risk of pathogen exposure and other impacts). They are described in more detail following Figure A-1.

Influences on an Integrated Risk Assessment and Risk Communications (RA/RC) System for Biosolids Land Application
Simple Expert Model

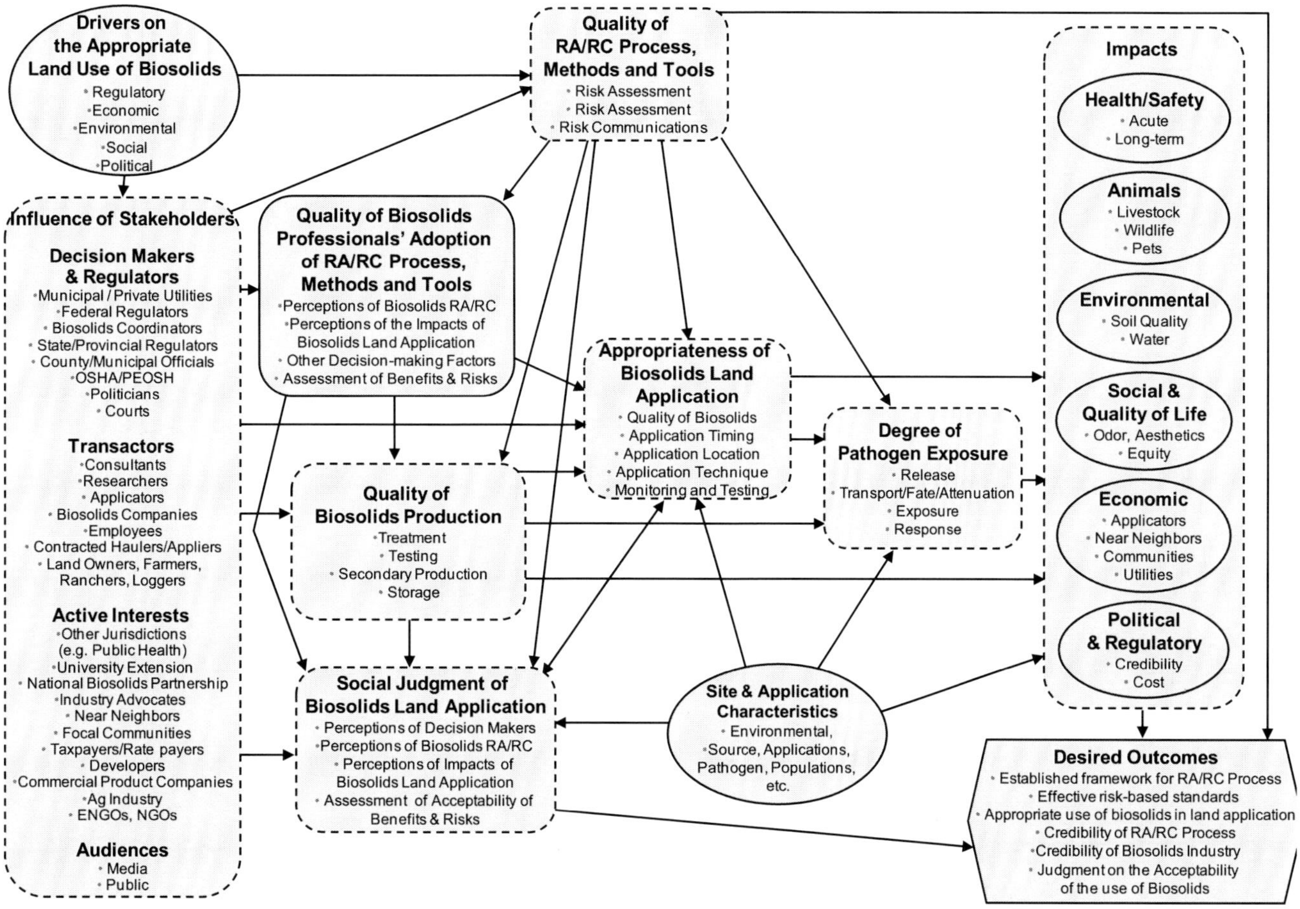

Figure A-1. Simple Expert Model.

A.2.3 Description of Model Nodes

A.2.3.1 Drivers on the Appropriate Land Use of Biosolids

Drivers represent the higher-level influences motivating the need or opportunity for a) biosolids land application and b) appropriate risk assessment and, ultimately, risk management of pathogens. As such, these drivers influence the decisions of the stakeholders regarding biosolids production and land application, as well as the quality of the RA/RC process, methods and tools. *Drivers* include:

- Regulatory drivers, such as the Clean Water Act and bans on ocean dumping of sewage sludge, which motivate the development of alternative approaches to manage wastewater and sewage sludge disposal or recycling. Another regulatory driver may be the quality and scope of current risk management standards and regulations, which may motivate the development of more systemic processes to understand and address risks of biosolids land application.

- Economic drivers, such as the costs of landfill disposal, which may motivate the development of alternative approaches to manage sewage sludge, on the one hand, and the positive economic impacts that may result from biosolids land application used as an alternative land supplement, on the other.

- Environmental drivers, such as recognition of the potential benefits of recycling nutrients, on the one hand, and the potential environmental and ecological risks posed by biosolids land application, on the other.

- Societal drivers, such as the increasing degree of public participation in community-level decision making, and legacy issues (e.g., salmonella contamination in fresh produce) that increase the public's level of concern about pathogen risks.

- Political drivers, such as the desire of a community to demonstrate leadership in environmental stewardship, or legacy issues related to biosolids land application that influence community leaders' judgments of the acceptability of biosolids land application projects.

A.2.3.2 Influence of Stakeholders on Risk Management of Biosolids Land Application

The Model depicts the range of stakeholders who may affect or be affected by biosolids land application projects.

These stakeholders can influence the system in a number of ways, for example: by advancing scientific understanding, establishing regulations, monitoring enforcement, setting budgets, and communicating to the public. These stakeholders are influenced by the *System Drivers* to varying degrees. For example:

- Municipal utilities' decisions in general may be influenced by regulatory drivers and economic drivers stemming from the cost of landfill disposal of sewage sludge. Some utilities may also recognize biosolids as a marketable product with potential economic benefits.

- Stakeholders within the biosolids profession may be driven more by their perceptions of the economic opportunities for developing marketable products.

- Environmental Non-governmental Organizations' (ENGOs') decisions within the system (e.g., to organize in an effort to influence decisions at a local level) may be primarily driven by their perceptions of the environmental and social drivers.

The stakeholders are categorized (*Decision Maker/Regulator, Transactor, Active Interest* and *Audience*) to indicate differing degrees to which they can affect or be affected by biosolids land application. The range of stakeholders and their roles in the system are depicted in **Sub-model A** and described in more detail in Section A.3.1.

A.2.3.3 Quality of RA/RC Process, Methods and Tools

The *Quality of RA/RC Process, Methods and Tools* is the degree to which the risk assessment and risk communications tools address the pathogen risk management needs for biosolids land application. The process developed for the purposes of Research Challenge is modeled after the Canadian Standards Q850[3] (described in Chapter 1.0 of this report). Among the methods and tools to support the process are the pathogen risk assessment tool and supporting communications, and the strategic risk communications strategy, plans and messages that serve as the deliverables of this Research Challenge. The *Quality of RA/RC Process, Methods and Tools* influences a number of other variables in this system, including:

- *The Quality of Biosolids Professionals' Adoption of RA/RC Process, Methods and Tools.* For example, adoption of the risk assessment tool or risk communications plans by utilities can be influenced by stakeholders' judgment of the quality of the tool, in terms of its accuracy, accessibility, and relevance.
- *The Quality of Biosolids Production, the Appropriateness of Biosolids Land Application and the Degree of Pathogen Exposure.* For example, the accuracy of the risk assessment tool may influence the appropriateness of decisions that are informed by the tool.
- *Social Judgment of Biosolids Land Application.* For example, through the quality of communications to community members.
- *Desired Outcomes.* For example, by directly influencing whether there is an established framework for RA/RC process or the effectiveness of risk-based standards.

The *Quality of RA/RC Process, Methods and Tools* is influenced by *Drivers,* for example, through bans on ocean dumping which increases the opportunity to advance acceptable land use of biosolids, through current scientific understanding (or lack thereof) of the risks of pathogens in biosolids land application, or through increased public discourse on the potential risks. This quality is also influenced by a range of *Influences of Stakeholders*, for example, through federal, state, and local regulations or through the quality of researchers' methods and tools for conducting risk assessment (e.g., sampling techniques). The process comprising *RA/RC Process, Methods and Tools* and the range of influences on the quality of production are depicted in Sub-model B and described in more detail in Section A.3.2.

---

[3] *Decision Partners' Principals Sarah Thorne and Gordon Butte presented this process to WERF at a meeting in October 2007. It is the base of the integrated approach currently underway on the Biosolids Challenge. Sarah was a coauthor of Q850, and Gordon was contributing member of the Technical Committee that developed this process. Sarah has recently been asked to serve on the new CSA Technical Committee responsible for upgrading Q850, based on experience, best practices and the evolving sciences of risk communications and risk management.*

A.2.3.4 Quality of Biosolids Production

The *Quality of Biosolids Production* represents the treatment of sewage sludge and production of biosolids for land application. It includes several processes, such as treatment, de-watering, odor control, monitoring and testing the product, and on-site storage. The *Quality of Biosolids Production* is influenced by a range of *Influences from Stakeholders*, the *Quality of Biosolids RA/RC Processes, Methods and Tools,* and *Site and Application Characteristics.* The process comprising *Biosolids Production,* the range of influences on the quality of production, and the range of impacts that may result from biosolids production is depicted in Sub-model C and described in more detail in Section A.3.3.

A.2.3.5 Appropriateness of Biosolids Land Application

The *Appropriateness of Biosolids Land Application* is the quality of biosolids land application as measured the by quality of the product applied to land, the appropriateness of the timing, location, and application of the biosolids, and the quality of monitoring and testing of the soil after the biosolids land application. Processes included are the selection of land application sites, hauling biosolids from the utility storage to the sites, application, monitoring testing, posting signs, and returning to land use after application. The *Appropriateness of Land Application* is influenced by:

- ♦ A range of *Influences from Stakeholders*; for example, regulations set by federal, state, and local government officials, as well as landowners' decisions to apply to be a potential biosolids land application site.
- ♦ The *Quality of Biosolids Production,* which will influence the characteristics of the biosolids end product, such as the moisture content, nutrient levels, pathogen levels, and odor. The biosolids end product (e.g., its moisture content) also influences decisions on how biosolids are transported, where and when they can be applied, and how they are applied.
- ♦ The *Quality of Biosolids RA/RC Process, Methods and Tools*, designed to support decisions on biosolids land application, and the *Quality of Biosolids Professionals' Adoption of Biosolids RA/RC Process, Method and Tools.*
- ♦ *Social Judgment of Biosolids Land Application,* which can influence such things as the selection of land application sites and transportation routes. Social judgment can ultimately influence whether biosolids projects are even allowed by community officials in the local community.
- ♦ A range of *Site/Application Characteristics*, such as environmental factors, community factors (e.g., a large metropolitan setting).

The process comprising biosolids land application, the range of influences on the quality of production, and the range of impacts that may result from biosolids production are depicted in Sub-model D and described in more detail in Section A.3.4.

A.2.3.6 Degree of Pathogen Exposure

The *Degree of Pathogen Exposure* is the potential environmental (land, air, water, plants) contamination and/or animal and human exposure to pathogens. It includes the release, transport, fate, and attenuation of the pathogens, as well as human exposure factors such as individual sensitivity and dose response. It also includes the quality of monitoring for such exposure and response to possible contamination. The *Degree of Pathogen Exposure* is influenced by:

- A range of *Influences from Stakeholders*; for example, landowners' decisions about allowing access to the land after biosolids land application, public health officials' decisions to step in and take action in response to a potential incident, and decisions by the media and others to communicate about possible incidents.

- The *Quality of Biosolids Production* and *Appropriateness of Biosolids Land Application*, which will influence the levels, fate, and transport of pathogens in the land.

The range of influences on the degree of pathogen exposure and the range of potential resulting impacts are depicted in Sub-model E, and described in more detail in Section A.3.5.

A.2.3.7 Impacts of Biosolids Land Application

Impacts of Biosolids Land Application describe the range of impacts from biosolids land application, including economic, health and safety, environmental, social, and regulatory impacts. The range of impacts is depicted in aggregate in Sub-model F and described in more detail in Section A.3.6.

A.2.3.8 Social Judgment of Biosolids Land Application

Social Judgment of Biosolids Land Application, which describes the stakeholders' assessment of the benefits and risks of biosolids land application, and their ultimate decisions on whether to support biosolids land application. It includes stakeholders' perceptions of a) the key decision makers in biosolids land application projects (e.g., the local municipal utility), b) the quality of biosolids risk assessment and risk communications, and c) the impacts of biosolids land application.

Social Judgment of Biosolids Land Application is influenced by:

- *The RA/RC Process, Methods and Tools* and the *Quality of Biosolids Professionals' Adoption of RA/RC Process, Methods and Tools,* including risk communications designed to address stakeholder interests and priorities.

- The *Quality of Biosolids Production*; for example, the level of pathogen removal and odor removal.

- The *Appropriateness of Biosolids Land Application*, particularly those decisions related to transport and land application; for example, decisions about transport routes, noise levels, and application timing may all influence how these stakeholders perceive the benefits and risks of biosolids land application.

Social Judgment of Biosolids Land Application also influences the *Appropriateness of Biosolids Land Application* and ultimately the *Desired Outcomes.* For example, landowners and neighbors' judgments can influence such things as the selection of application sites and transportation routes, timing of the application, etc. Social judgment can ultimately influence whether biosolids can be applied in specific locations. More details on *Social Judgment of Biosolids Land Application* are depicted in Sub-model G and described in Section A.3.7.

A.2.3.9 Site and Application Characteristics

The Model also identifies the range of site- and application-characteristics that may influence decisions related to biosolids land application. A clear finding from the expert model development is the need to consider the potential differences among biosolids land application projects, including environmental characteristics, water sources, and community populations. For example, climate conditions in the Southwest may influence decisions differently from those in application projects in cooler or rainy climates. As another example, *Social Judgment of Biosolids Land Application* by community citizens within states like Oklahoma is expected to differ from judgments of citizens in other communities where agriculture is not a significant economic source.

A.2.3.10 Desired Outcomes

Desired Outcomes represent the ultimate end-goals of the RA/RC system – to improve decision making on the sustainable use of biosolids in land application. *Desired Outcomes* include:

- An established framework for biosolids pathogens risk assessment and risk communications, that guides effective assessment and, ultimately, mitigation and/or management of pathogens risk in biosolids land application. One goal of this Research Challenge is to inform the development and implementation of such a framework.
- Effective risk-based standards for pathogens in biosolids land application, stemming from this framework. Another goal of this Research Challenge is to provide policy makers with robust theoretical and empirical data to support the development of these standards.

Also included are broader desired outcomes of an integrated risk assessment and risk communications system, including:

- Appropriate use of biosolids in land application that is guided by an analytical framework and supported by effective risk-based standards for pathogen risk assessment and risk management.
- Stakeholder judgment of the credibility of biosolids risk assessment and risk communications processes, which includes their trust in decision makers and confidence in the effectiveness of these processes.
- Stakeholder judgment of the credibility of the biosolids profession, including trust in the people in the profession, and their ability to ensure the safe and appropriate use of biosolids land application.
- Stakeholder judgment of the acceptability of biosolids land application, stemming from their judgment of the credibility of risk assessment and risk communications processes, and their own assessments of the benefits and risks of biosolids land application.

A.3 Detailed Expert Sub-Models

As mentioned, the **Simple Expert Model** describes a complex system related to biosolids land application projects. The **Detailed Expert Sub-Models** describe some of the variables, or "nodes" of the system in greater detail. The Model has been divided into six sub-models that focus on particular nodes of the **Simple Expert Model**. These are as follows:

- **Detailed Expert Sub-Model A: Stakeholder Map of Biosolids Land Application** identifies the range of stakeholders related to biosolids land application and biosolids risk management. Stakeholders are grouped into categories based on the degree to which they affect or can be affected by biosolids land application.

- **Detailed Expert Sub-Model B** focuses on the *Quality of RA/RC Process, Methods and Tools* to support pathogen risk assessment, risk communications, and ultimately risk management of biosolids land application.
- **Detailed Expert Sub-Model C** focuses on the *Influences on the Quality of Biosolids Production* and depicts the economic, social, health, and regulatory impacts that may stem directly from biosolids production (i.e., regardless of whether those biosolids are land applied).
- **Detailed Expert Sub-Model D** depicts the *Influences on the Appropriateness of Biosolids Land Application* and highlights potential economic, health, social, environmental, and regulatory outcomes of biosolids transport and application.
- **Detailed Expert Sub-Model E** focuses on the *Influences on the Degree of Pathogen Exposure* and depicts environmental, health, economic, social, and regulatory impacts of pathogen exposure.
- **Detailed Expert Sub-Model F** focuses on the *Impacts of Biosolids Land Application* and depicts the range of health and safety impacts, animal impacts, environmental impacts, social and quality of life impacts, economic impacts, and political and regulatory impacts.
- **Detailed Expert Sub-Model G** focuses on the *Influences on Social Judgment of Biosolids Land Application* and depicts the influences on Individuals' Mental Models and their judgments and assessment of the benefits and risks of biosolids production and land application.

A.3.1 Detailed Expert Sub-Model A: Stakeholder Map of Biosolids Land Application

The first detailed Sub-model, the **Stakeholder Map** (Figure A-2), is an effective and efficient tool to identify and categorize stakeholders. Biosolids land application is at the center of a complex system, involving many different stakeholders: water utilities who are essential to managing the treatment of wastewater into safe and effective biosolids; risk management researchers and experts who influence risk management policies and practices; regulators who set standards for land application practices; utility employees, contractors, and other land appliers who make, transport and/or apply biosolids; municipal officials who control budgets; and community members, who give input as to the acceptability of its use in their communities.

Decision Partners worked with WERF and members of the Biosolids Research Challenge Team to develop the **Stakeholder Map of Biosolids Land Application**. At the center of the Map is the risk challenge: public judgment of the acceptability of biosolids land application. Each of the groups or individuals identified has a relationship with a biosolids land application project that is more or less direct, depending on its proximity to the center of the concentric circles. Thus, the closer they are situated to the center, the more likely they will have a direct "stake" in decisions related to biosolids land application.

LEGEND

Decision Makers / Regulators: Stakeholders who have direct decision-making authority to establish rules about biosolids and/or can influence policy decisions about biosolids land application.

Transactors: Stakeholders who have some sort of role or direct transaction (e.g., financial) in the decision-making process.

Active Interests: Individuals or groups who have a stake in the issue of biosolids land applications and risk management (e.g., can be directly impacted), but are not directly involved in the decision-making process.

Audience: Individuals or groups that may have an interest in the topic of biosolids land applications and risk management, but who are not directly affected by nor involved in the decision-making process.

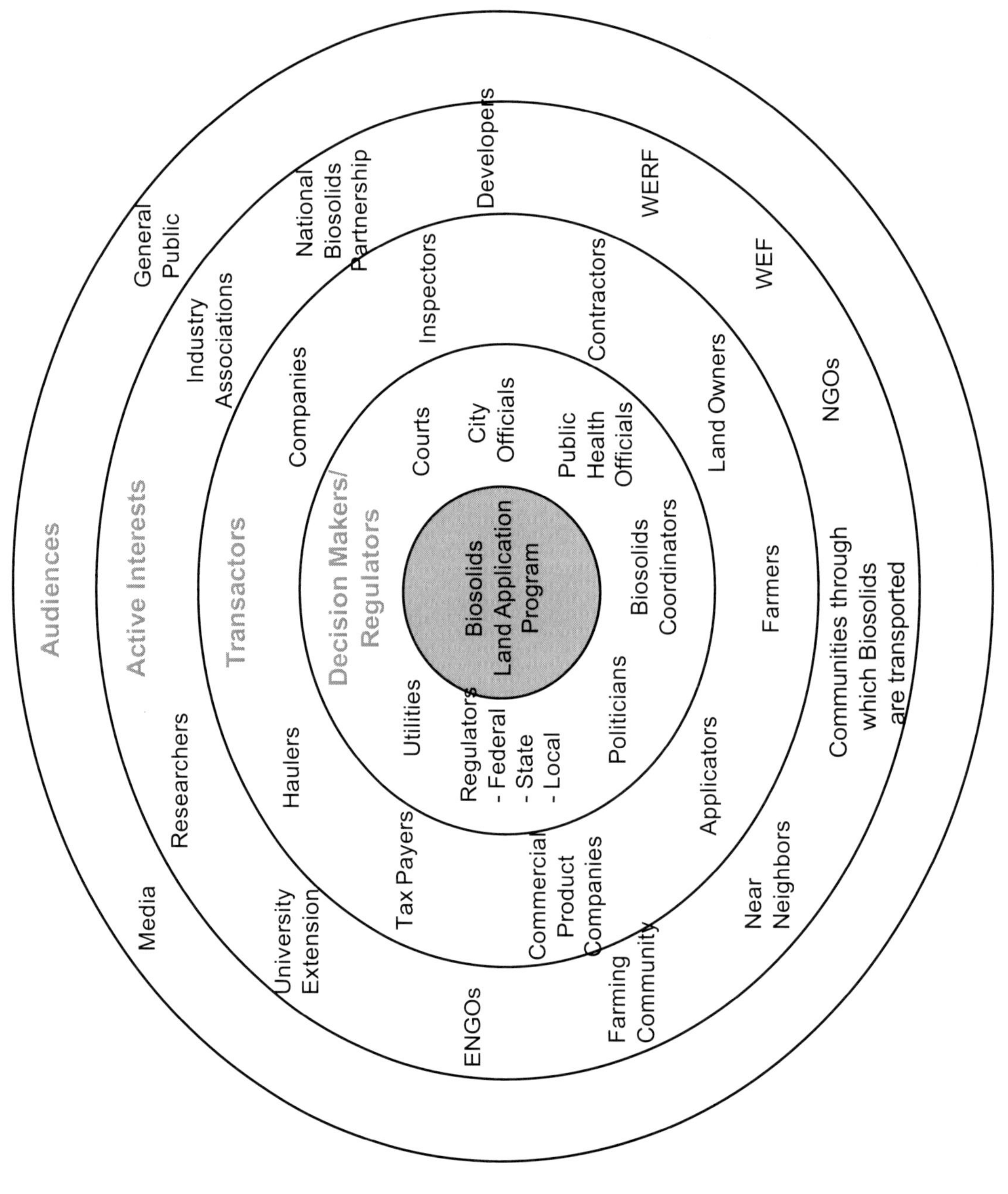

Figure A-2. Detailed Expert Sub-Model A.

A.3.1.1 Stakeholder Categories

The four categories of stakeholders shown on the Map are:

- *Decision Makers/Regulators*: are stakeholders who have direct decision-making authority to establish rules about biosolids and/or can influence policy decisions about biosolids land application. *Decision Makers/Regulators* can include agencies, departments, and/or jurisdictions that have a primary or shared role in regulating or managing the risk. Examples of *Decision Makers/Regulators* include:
  - *Utilities*, including municipal public works departments and private utilities, who typically are key decision makers in the local biosolids land application projects. Utilities' staff may make decisions about:
    - Whether to initiate biosolids land application projects.
    - How to treat, produce, store, and transport biosolids.
    - Where and how biosolids are applied.
    - Whether to use contractors to support biosolids land application. If contractors are used, utilities' staff typically oversee the contractors' work and performance.
    - Communicating with stakeholders, including community officials, landowners, and citizens about biosolids land application.
  - *Federal, State, and Local Government Regulators;* for example, the U.S. Environmental Protection Agency. *Regulators* are primarily responsible for promulgating regulations, setting standards, and establishing policy on biosolids land application.
  - *Biosolids coordinators*, who have regulatory oversight of biosolids land application projects at a regional or state level. Biosolids coordinators in some states have the responsibility to permit land application sites.
  - *County/Municipal Officials* including mayors and municipal council members. These decision makers typically control or influence utility budgets.
  - *Courts*, which may exert influence on the use of biosolids land application through their legal decisions.
  - Occupational Safety and Health Agency (OSHA) and Public Employees Occupational Safety and Health Unit (PEOSH).
- *Transactors:* Stakeholders who are most directly affected by the decisions regarding biosolids land application and risk management and have some sort of role or transaction in the decision-making process. Examples of *Transactors* include:
  - Utility employees, who are responsible for performing or overseeing biosolids land application. Utility employees are especially important to consider when developing and implementing risk assessment and risk communications processes, methods and tools. While not typically decision makers in the sense of setting standards for managing and communicating risk, they can often directly affect how the standards are implemented. For example, they can affect whether

requirements for minimum application distance or maximum rates are met. They may also have direct interaction with other stakeholders in the community, such as near neighbors.

- Contracted haulers or land appliers, who are hired by utilities to manage or support specific operations, such as transporting biosolids or land applying. Similar to utility employees, they have a direct influence on biosolids land application risk management and risk communication.

- Landowners, ranchers, loggers, and farmers, who own, lease, or sell land for biosolids land application. These stakeholders may receive payment or may be required to pay for the biosolids land application. They may experience direct economic benefits, such as increased crop yield. Similar to utility employees and contracted haulers and land appliers, these stakeholders also influence biosolids land application risk management and risk communication. For example, they may have direct control over when their animals return to grazing after applications; they may be required to do some preparatory work in the field; and they may directly communicate with neighbors and other landowners in the community.

- Biosolids companies, who contract with utilities to oversee biosolids projects or market biosolids products, such as compost and other soil amendments.

- Consultants, who support regulators and/or the biosolids profession by conducting risk assessment, developing production and applications processes, advancing policies, and/or guiding communication efforts.

- Researchers, who support regulators and the biosolids profession , by advancing scientific understanding of biosolids risk assessment, risk communications and risk management. Researchers may receive funding, access to data, and other direct benefits.

♦ *Active Interests*: Individuals or groups who have a stake in the issue of biosolids land application, but are not directly involved in the decision-making process. Examples of *Active Interests* include:

  - Land developers, who may develop on land sites on land that a) has been previously applied with biosolids, and/or b) is near current biosolids land application. These stakeholders may be particularly influential in cities experiencing expanding suburban development into predominately rural areas.

  - Agriculture industry, at many levels, including meat or grain processors, fertilizer companies, and others.

  - Environmental non-governmental organizations (ENGOs) and other non-government organizations (NGOs).

  - Other regulatory jurisdictions (e.g., public health), who may not have direct policy or regulatory responsibility, but who have a stake in ensuring appropriate biosolids land application. These jurisdictions may play a more significant role if, for example, contamination incidents occur or are suspected (e.g., truck spills, water contamination).

- Near neighbors of biosolids land application, and the Communities Through Which Biosolids are Transported, are the people who may be, or may perceive themselves to be, affected by biosolids land application without direct benefits.
- Taxpayers and/or rate payers, who may be affected by economic impacts from biosolids land application (or, conversely, the alternatives to land application, such as landfill disposal, which may be more costly).
- WERF, WEF, biosolids associations (e.g., National Biosolids Partnership) and other organizations whose missions are to support the biosolids profession to ensure the appropriate use of biosolids land application.
- University extension, which may have a significant role in liaising with researchers, regulators, utilities, and land owners to ensure the appropriateness of biosolids land application.
- Commercial product companies, who market products that may contain biosolids.

♦ *Audiences*: Individuals or groups that may have an interest in the topic of biosolids land application, but who are not directly affected by nor involved in the decision-making process. Examples of *Audiences* include the media and the general public.

This Stakeholder Map is intended to be generic, for the biosolids profession as a whole. In any given community, the Map becomes dynamic and stakeholders can move into different categories depending on how they affect or are affected by decisions over time. For example, a landowner may first be Active Interests (*near neighbors)* and become Transactors (*landowners)* if they offer their land as a biosolids land application site.

A.3.2 Detailed Expert Sub-Model B: Quality of RM/RC Process, Methods, and Tools

Sub-model B in Figure A-3 focuses on the *Quality of RA/RC Process, Methods and Tools* for pathogen risk assessment and risk communications to support effective risk management. The process developed for the purposes of Research Challenge, illustrated in the model, includes iterative processes of *Initiation, Risk Assessment, Risk Evaluation, Risk Control* and *Action/Monitoring.* Among the methods and tools to support the process are the pathogen risk assessment tool and supporting communications, and the strategic risk communications strategy, plans, and messages that serve as the deliverables of this Research Challenge. Each step of the process has both technical and risk communications components. This process is modeled after the Canadian Standard Associations' *Q850 Risk Management: Guideline for Decision-Makers*[4] (described in Chapter 1.0 of this Report). This process is also aligned with integrated risk management and risk communications processes that have been adopted by Health Canada, the U.S. Presidential/ Congressional Commission on Risk Assessment and Risk Management Process, the Australian/ New Zealand Risk Management Standard, and is currently being considered for inclusion in the International Standards Organization (ISO) 31000 Guideline on Risk Management.

---

[4] *Decision Partners' Principals Sarah Thorne and Gordon Butte presented this process to WERF at a meeting in October 2007. It is the base of the integrated approach currently underway on the Biosolids Challenge. Sarah was a coauthor of Q850, and Gordon was contributing member of the Technical Committee that developed this process. Sarah has recently been asked to serve on the new CSA Technical Committee responsible for upgrading Q850, based on experience, best practices and the evolving sciences of risk communications and risk management.*

A.3.2.1 Description of Model Nodes

- The *Quality of RA/RC Process, Methods and Tools* influences a number of other variables in this system, the *Quality of the Biosolids Professionals' Adoption of RA/RC Process, Methods and Tools* and *Social Judgment of Biosolids Land Application*, for example, through the quality of communications to community members.
- The *Quality of RA/RC Process, Methods and Tools* is influenced by a number of factors, including *Drivers*, for example through bans on ocean dumping which influences the Need for Wastewater Management, through current scientific understanding (or lack thereof) of the risks of pathogens in biosolids land application, or through increased public discourse on the potential risks. This quality is also influenced by a range of Influences of Stakeholders, the Coordination of Key Decision Makers and the quality Risk Assessment Tools for conducting risk assessment (e.g., sampling techniques).

Detailed Expert Sub-Model B: Influences on the Quality of WERF RA/RC Process, Methods and Tools

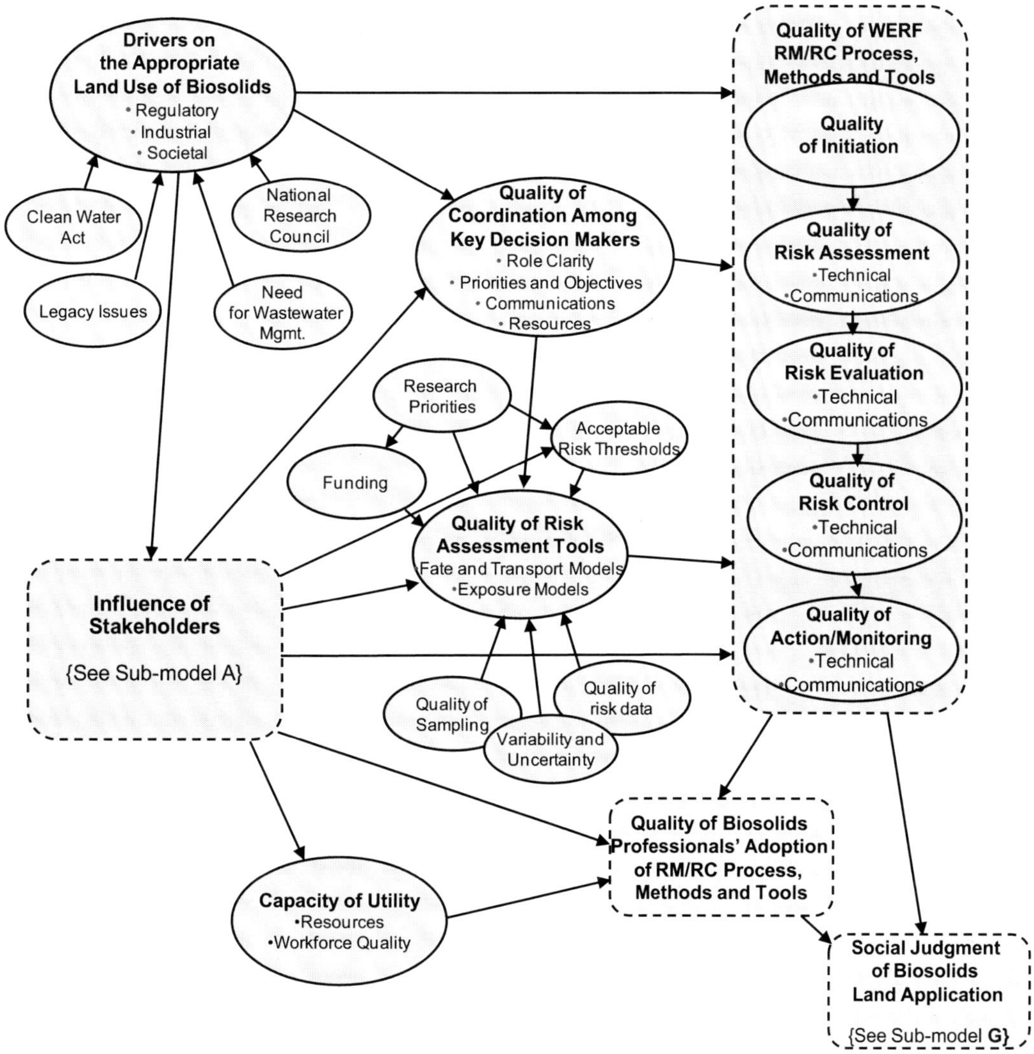

Figure A-3. Detailed Expert Sub-Model B.

A.3.3 Detailed Expert Sub-Model C: Quality of Biosolids Production

Detailed Expert Sub-model C in Figure A-4 focuses on the *Influences on the Quality of Biosolids Production* and details the economic, social, health, and regulatory impacts that may result from biosolids production (i.e., regardless of whether those biosolids are land applied). The Sub-model indicates different types of *Impacts* of the system with gray shaded triangles.

A.3.3.1 Description of Model Nodes

Near the center of the Sub-model is a series of nodes that comprise the different elements of biosolids production. These include:

- Treatment of sewage sludge for the purposes of wastewater treatment, pathogen reduction, and odor control. Several treatment methods exist, such as biological or chemical stabilization (e.g., anaerobic, aerobic, lime stabilization), thermal processes, composting, and full pasteurization.
- Testing to ensure safe levels of pathogens and other materials (e.g., nitrogen content). Testing procedures may vary widely among utilities, including those who do not have formal testing procedures.
- Secondary Production, including de-watering and further odor control. These processes may have a significant impact on when, where, and how biosolids are land applied; for example, whether as a semi-solid ("cake") or as a liquid.
- On-site Storage, depending on the form of the biosolids (e.g., cake or liquid). Storage may require facilities such as lagoons or holding beds, as well as additional machinery to transport and move the biosolids to and from on-site storage. Storage may also require protocols and procedures to ensure appropriate maintenance and removal.

Any processes beyond on-site storage are modeled in Sub-model D: *Appropriateness of Biosolids Land Application.*

The Sub-model also describes a number of influences on how biosolids are produced, which are shown on the left-hand side of the diagram. These include:

- *U.S. EPA Part 503 Regulations*: Standards for the Use or Disposal of Sewage Sludge, which set limits for pathogens and other pollutants in treated sewage sludge, defines operational processes to achieve those limits, establishes requirements for biosolids land application, and sets standards for monitoring and reporting.
- *Other State and Local Regulations*, which may be stricter on biosolids land application than the U.S. EPA Part 503 Regulations. For example, Local Regulations may allow land application of only Class A biosolids.
- A number of *Utility Factors*, such as facility size, location(s), staff, resources, etc., as well as a number of *Source Water Factors*, such as volume, base level of pathogens and other contaminants, etc.
- A number of *Application Factors*, such as whether the applied biosolids are Class A or Class B typical hauling distances, types of land, etc.
- A number of *Environmental Factors*, such as typical climate conditions (e.g., expected rainfall, temperature), topography, etc.

Influences on an Integrated Risk Assessment and Risk Communications (RA/RC) System for Biosolids Land Application
Detailed Expert Sub-Model C: Influences on the Quality of Biosolids Production

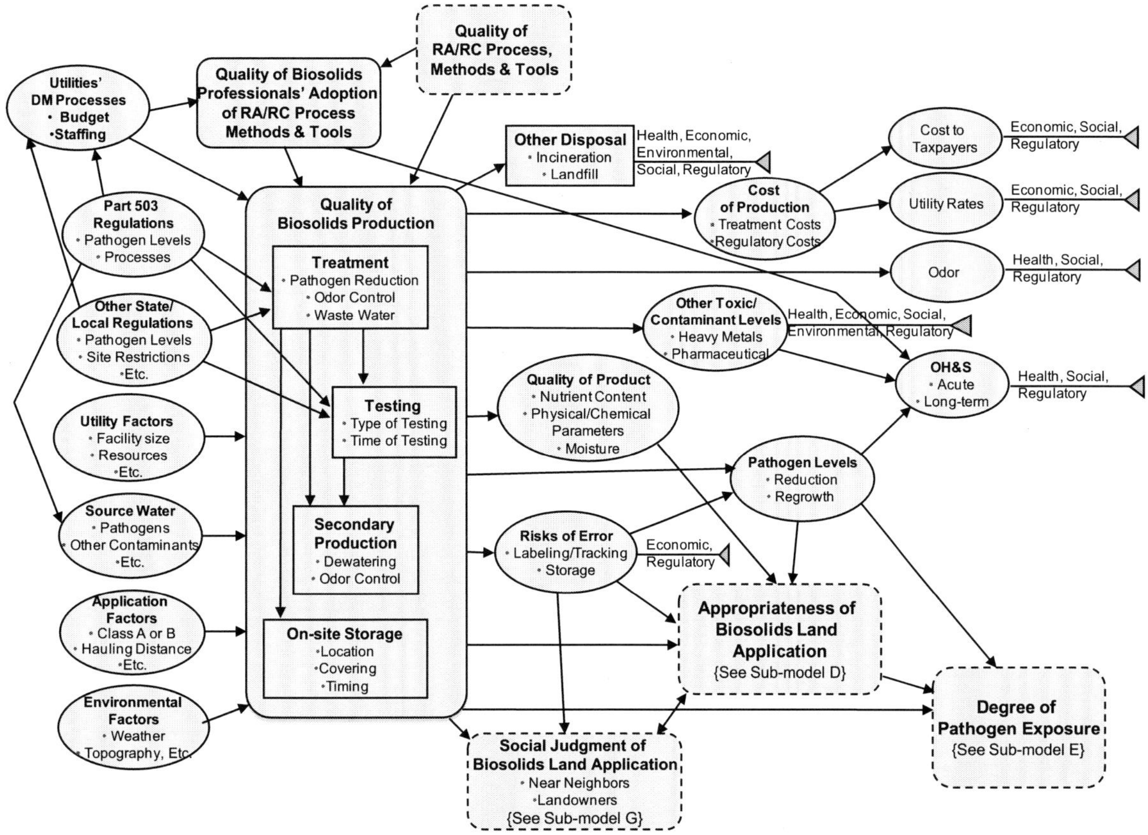

Figure A-4. Detailed Expert Sub-Model C.

Sub-Model C on the previous page also illustrates how the *Quality of Biosolids Production* can influence other variables in the system. The Sub-Model shows how the *Quality of Biosolids Production* can influence the *Appropriateness of Biosolids Land Application* directly (e.g., whether cake or liquid is produced). It can also affect it indirectly through the *Quality of Product,* in terms of nutrient content, moisture content, *Degree of Pathogen Levels*, which may or may not be known through testing, and through *Risks of Error*, such as tracking and storage errors. The *Quality of Biosolids Production* can also have an impact on *Social Judgment of Biosolids Land Application* directly (e.g., through stakeholders' perceptions on the appropriateness of treatment and storage processes) and indirectly through *Risks of Error,* such as an on-site contamination incident. Other impacts of biosolids production include:

- Utilities' decisions, shown by a square decision box, on whether *Other Disposal* methods such as landfills are required. Decisions about *Other Disposal* will influence a number of economic, health, environment, social, and regulatory impacts.
- The *Cost of Production*, which can vary significantly depending on the types of treatment processes used. The *Cost of Production* can influence *Utility Rates* and *Cost to Taxpayers.*
- The presence of *Other Toxic/Contaminants Levels* such as heavy metals and trace organic compounds.
- *Odor* of the biosolids produced, as well as odor generated during biosolids production and storage.
- *Occupational Health and Safety* (OH&S) of employees and contractors who work with sewage sludge and biosolids during the production and on-site storage phases. *OH&S* can be affected by *Pathogen Levels* as well as *Other Toxic/Contamination Levels.*

A.3.4 Detailed Expert Sub-Model D: Influences on the Appropriateness of Biosolids Land Application

Detailed Expert Sub-Model D in Figure A-5 focuses on the *Influences on the Appropriateness of Biosolids Land Application* and depicts the economic, social, health, and regulatory impacts of biosolids application.

A.3.4.1 Description of Nodes

Near the center of the Sub-Model is a series of nodes that comprise biosolids land application. These nodes describe the key processes involved in transport, application, and post-application of biosolids. They include:

- *Applicator Selection*, which accounts for the selection and oversight of employees and contractors to transport and apply biosolids.
- *Site Selection*, which accounts for the selection of biosolids land application sites and oversight of the preparatory work on those sites.
- *Hauling*, which is the transport of biosolids from the biosolids production facility to the land application site. It includes selection of transport routes, maintenance and operation of transport vehicles, and protocols for dumping biosolids at the site in advance of actual application. From a communications perspective, it includes any

use of signs, labels, notifications, and other communications related to biosolids transport.

- *Storage* describes any use of off-site storage facilities between production and use, if applicable.
- *Application*, which describes the process of applying biosolids and any additional processes used to work the applied materials into the land (e.g., harrowing or plowing). This includes determining when biosolids can be applied (e.g., taking into account weather conditions) and rates of application (e.g., using agronomic rates to account for pre-existing nutrients in the soil). It may also include communication plans and messages for staff and contractors, and operations and maintenance procedures for application equipment. Additionally, it may include any use of signs, labels, notifications, and other communications about biosolids land application.
- *Post Application Processes*, which accounts for any requirements or procedures following a biosolids application, including allowing land use post-application, and monitoring and testing pollutant and nutrient levels.

The Sub-Model also describes a number of influences on the quality of biosolids land application, which are shown on the left hand side. These include:

- *Applicator Permit*, which is influenced by the *Degree of Competition for Sites*. The degree of competition varies from one community to any another and depends on a number of factors, such as typical land use. Agricultural communities, for example, may have a high demand for biosolids land application.
- *Site and Application Factors*, such as transport distances, land size, topography, and the location of residences, roads, water sources, etc.
- *U.S. EPA Part 503 Regulations* and standards for the use or disposal of sewage sludge, which stipulate various conditions, such as minimum distances from residences or water supplies and permitted weather conditions for applying biosolids.
- *Other State and Local Regulations*, which may be stricter on biosolids land application than the U.S. EPA Part 503 standards. For example, local regulations may require longer time before permitting grazing or use of land after biosolids land application. These regulations may also influence the *Applicator Permitting* process.
- The *Quality of Biosolids Production.* As described in Sub-Model C, the *Quality of Biosolids Production* through the *Quality of Product, Risks of Error* and *Pathogen Levels from Biosolids Productions.*
- A number of *Environmental Factors*, such as climate (e.g., rainy or arid climates, winter climates), weather (wind, humidity, precipitation, temperature), topography, soil conditions, and nutrient.
- *Social Judgment of Biosolids Land Application*, which can influence the *Appropriateness of Biosolids Land Application* a number of ways; for example, the selection of land application sites and transportation routes. *Social Judgment* can ultimately influence the feasibility and acceptability of biosolids land application projects in local communities.

Finally, the Sub-Model illustrates how the *Appropriateness of Biosolids Land Application* can influence other variables in the system, including:

- *Quality of the Soil Amendment*, which is the beneficial supplementation of nutrients and other elements to the land, and the minimization of negative ones.
- *Risks of Error*, which includes the risk of biosolids spills, applying to the wrong location, and inciting cross contamination of pathogens.
- *Degree of Pathogen Exposure*, which can be influenced directly by biosolids land application or indirectly, through *Pathogen Levels*, and *Exposure Risk* (e.g., the risk of coming in contact with contaminated biosolids through the inadequate quality of employees' personal protective equipment).
- A number of other factors, such as *Land Value, Marketability of Crops, Odor, Financial Impacts to Communities* and *Financial Impacts to Land Owners* (through *Direct Payments).*

Influences on an Integrated Risk Assessment and Risk Communications (RA/RC) System for Biosolids Land Application
Detailed Expert Sub-Model D: Influences on the Appropriateness of Biosolids Land Application

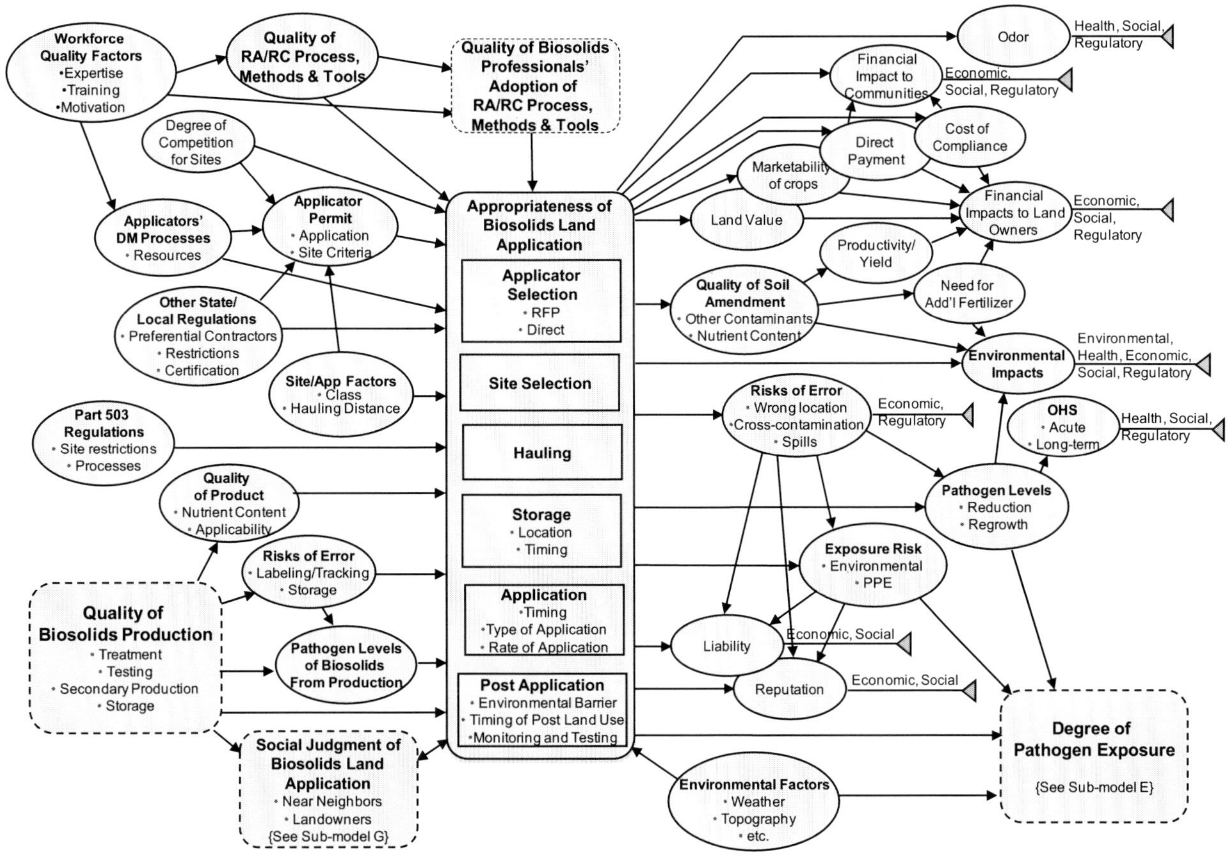

Figure A-5. Detailed Expert Sub-Model D.

A.3.5 Detailed Expert Sub-Model E: Influences on the Degree of Pathogen Exposure

Detailed Expert Sub-Model E on the next page focuses on the *Influences on the Degree of Pathogen Exposure* and details the economic, social, health, and regulatory impacts of pathogen exposure.

A.3.5.1 Description of Nodes

- *Degree of Pathogen Transport.* Factors that influence the *Degree of Pathogen Transport* include exposure routes (e.g., surface water, groundwater, soil type) and attenuation, or the weakening of pathogens over time.
- *Risks of Error*, including spills, wrong location, inappropriate application rates, and land use before the minimum 30-day period. *Risk of Errors* can influence *Pathogen Levels in Applied Biosolids*, and the *Degree of Pathogen Transport*. For example, cattle set to graze on biosolids-applied land too soon after application may have an increased risk of pathogen exposure.
- *Environmental Factors*, such as weather, topography, soil temperature, humidity, soil content, etc., which can influence pathogen exposure through the *Degree of Pathogen Transport* and influence specific *Pathogen Factors*.
- *Individual Factors*, which can increase the *Degree of Pathogen Exposure*, including factors that increase the likelihood of risk of exposure through human contact (e.g., employment).
- *Individual Sensitivity and Dose Response* of pathogen exposure, for example increased sensitivity by immuno-compromised individuals.
- *Health Outcomes and Markers*, which are the health impacts (e.g., acute illnesses) of pathogen exposure, and the symptoms that could influence the quality of detection of a pathogen incident.
- *Risk of Exposures or Health Impacts from Other Sources*, which accounts for other risks of pathogen exposure (e.g., animal-derived fertilizers).
- *Quality of Detection and Treatment* of pathogen exposure or its related health impacts.

Influences on an Integrated Risk Assessment and Risk Communications (RA/RC) System for Biosolids Land Application
Detailed Expert Sub-Model E: Influences on the Degree of Pathogen Exposure

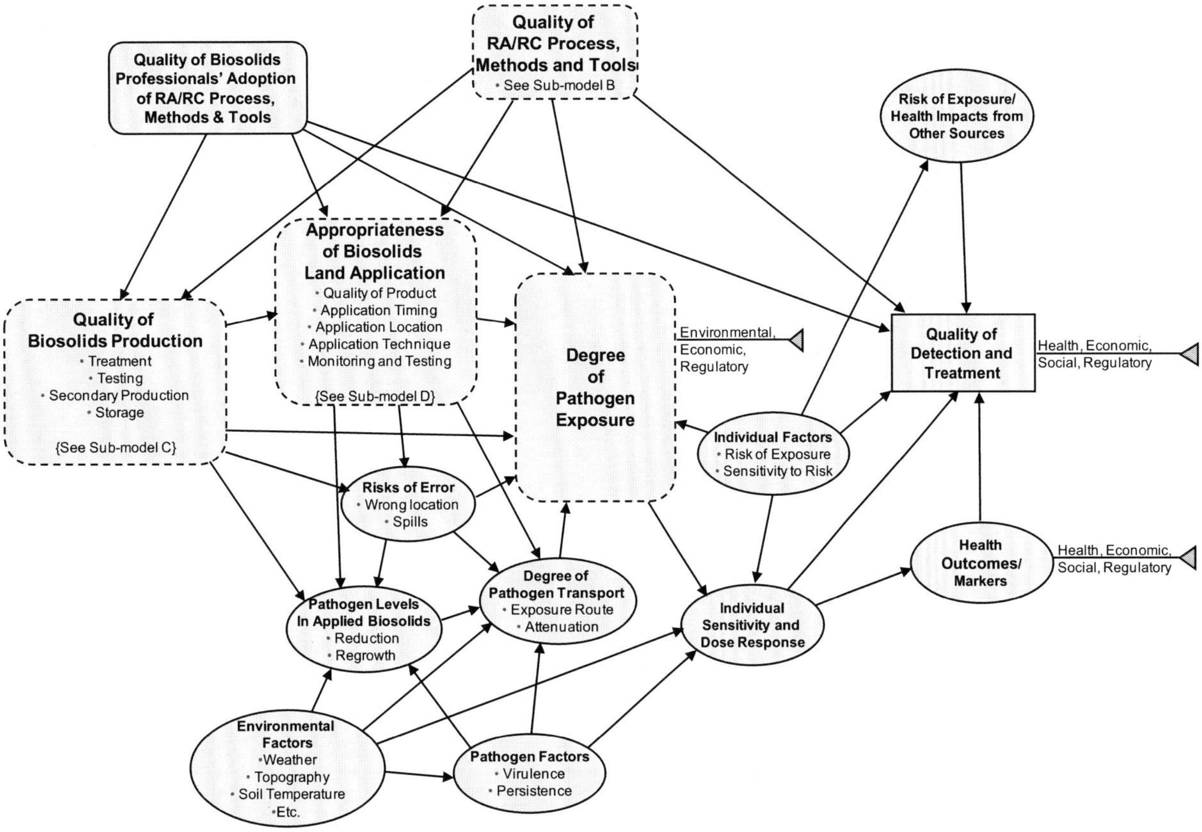

Figure A-6. Detailed Expert Sub-Model E.

A.3.6 Detailed Expert Sub-Model F: Impacts of Biosolids Land Application

Detailed Expert Sub-Model F in Figure A-7 provides a detailed set of possible *Impacts of Biosolids Land Application.*

A.3.6.1 Description of Nodes

- The *Health and Safety Impacts* include acute health impacts, long-term and/or cumulative impacts, impacts to children, accidents, occupational health impacts, and non-pathogen impacts, as well as odor and peace of mind. Odor and peace of mind also influence the *Social and Quality of Life Impacts*.
- *Animal Impacts* include impacts to pets, livestock, and wildlife. For example, farmers may need to keep their livestock away from treated areas for a period of time; however, wildlife may not be prevented from gaining access to those areas, and the impact of their exposure to those areas may be unknown.
- *Environmental Impacts* include ecological impacts as well as impacts to the groundwater, surface water, air, soil quality, and land use. For example, runoff from treated areas may affect the quality of the lakes, streams, and rivers in the area.
- *Social and Quality of Life Impacts* include land use, truck traffic, aesthetics, inconvenience, reputation, and equity issues. Odor and peace of mind also influence the *Social and Quality of Life Impacts*. For example, the amount of biosolids produced may not meet landowners' demand and therefore give rise to equity issues when landowners perceive unfair allocation of the biosolids. Conversely, residents of rural communities may perceive unfair distribution of biosolids from larger metropolitan areas.
- *Economic Impacts* include financial impacts to the community, landowners, contractors, and utilities. Examples of financial impacts to the community include increased taxes and utility rates, employment, remediation costs, and costs related to health impacts. Financial impacts to landowners include remediation costs, costs related to health impacts, land value, direct payment, and revenue increases due to crop yield. Financial impacts to contractors include increased revenue. Financial impacts to utilities include direct payment, revenue, liability, and RA/RC costs. Financial impacts to utilities would also directly influence the financial impacts to the community.
- *Political and Regulatory Impacts* include regulatory costs, political pressures and credibility. For example, the credibility of a town council that supports a biosolids land application program may be directly affected by the success or failure of the program.

Finally, the Sub-Model illustrates how several of the Impacts described above influence other elements in the system. For example, the *Environmental Impacts* influences both the *Animal Impacts* and *Health and Safety Impacts*, as well as on the *Social and Quality of Life Impacts*, *Economic Impacts* and the *Political and Regulatory Impacts*. Contaminated groundwater could have a negative impact on the health and safety of those living in the community and on the animals, as well as negative economic impacts due to increased remediation costs, potential decreases in land value, and loss of revenue for the contractors and utilities.

Influences on an Integrated Risk Assessment and Risk Communications (RA/RC) System for Biosolids Land Application
Detailed Expert Sub-Model F: Impacts of Biosolids Land Application

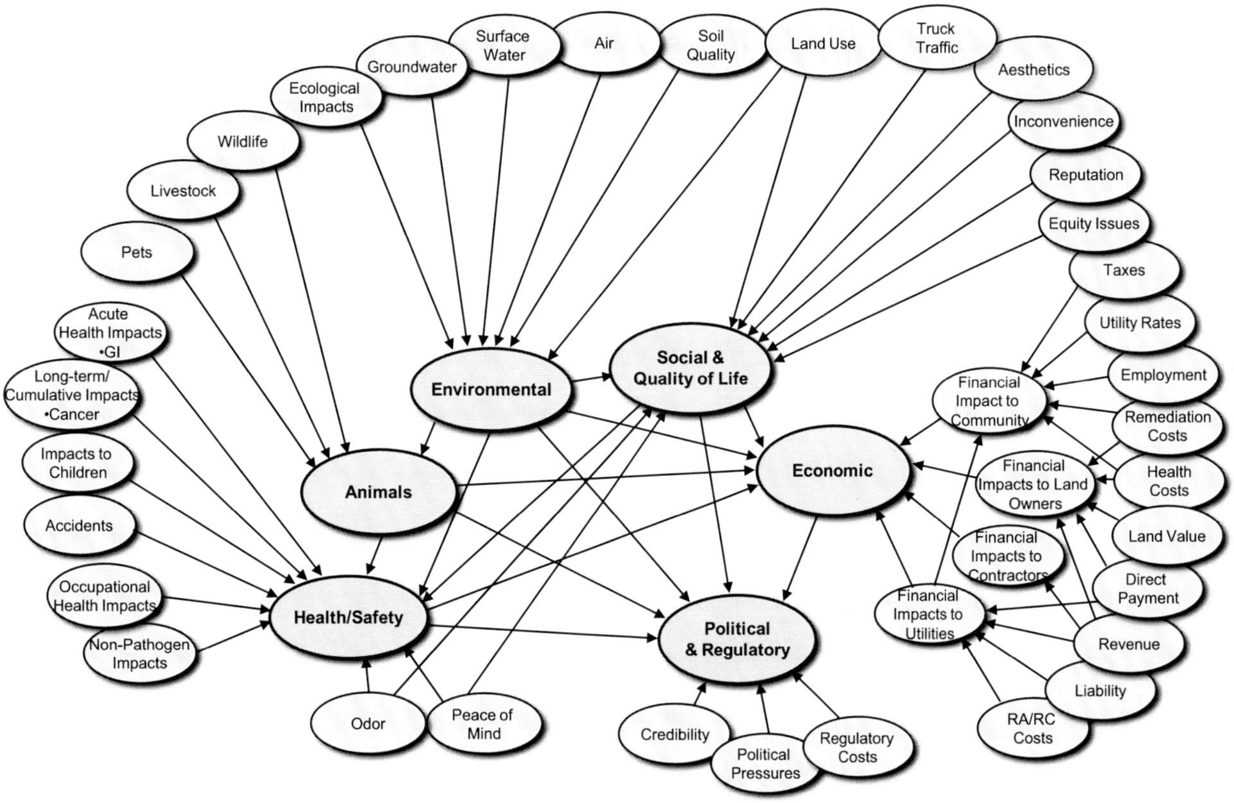

Figure A-7. Detailed Expert Sub-Model F.

A.3.7 Detailed Expert Sub-Model G: Influences on Social Judgment of Biosolids Land Application

Detailed Expert Sub-Model G on the next page describes experts' understanding, or hypotheses, of stakeholders' perceptions and assessment of the benefits and risks of biosolids land application and ultimate decisions to support such projects. For the purpose of this project (including the mental models research), the Sub-Model focus is on landowners' and near neighbors' judgments of the acceptability of local biosolids land application. However, the Sub-Model could also include the social judgments of community citizens, municipal officials, and the broader public.

Social Judgment of Biosolids Land Application can influence biosolids land application projects in a number of ways, and ultimately the *Desired Outcomes.* For example, landowners' and neighbors' judgments – and motivation to support or resist projects – can influence such factors as the selection of transportation routes and biosolids land application site, communications about biosolids and land use applications, etc. Social judgment can ultimately influence whether biosolids are a) socially acceptable and b) can be applied in specific locations.

The Sub-Model shows that an *Individual's Judgments of Biosolids Land Application* are influenced by the *Individual's Mental Model of Biosolids Land Application.* As described in the Introduction, a person's mental model is the complex web of deeply held beliefs that guide his or her judgments and decision making (Morgan et al., 2002).

A.3.7.1 Description of Nodes

- *Perceptions of Biosolids Professionals/Regulators* including their perceptions of the competency and trustworthiness of biosolids professionals (e.g., utilities), as well as the *Transparency of the Decision-Making Process* and the individual's *Degree of Control in Decision Making*. For example, neighbors may perceive that they have little control over whether biosolids are applied on adjacent land or that the decision has already been made.

- *Perceptions of Biosolids Land Application*, which include the individual's perceptions of production, land application, and the risk of pathogen exposure (depicted in Sub-Models C, D, and E, respectively). These perceptions may or may not be aligned with experts' understanding of these processes.

- *Perceptions of the Impacts of Biosolids Land Application*, as they were modeled in Sub-Model F. Again, Individual's perceptions may or may not be aligned with experts' understanding of these processes.

- The *Individual's Assessment of the Acceptability of the Benefits and Risks of Biosolids Land Application*, which represents how individuals weigh the benefits and risks of biosolids land application in their area in order to make a decision about their acceptability.

Influences on an Integrated Risk Assessment and Risk Communications (RA/RC) System for Biosolids Land Application
Detailed Expert Sub-Model G: Influences on Social Judgment of Biosolids Land Application

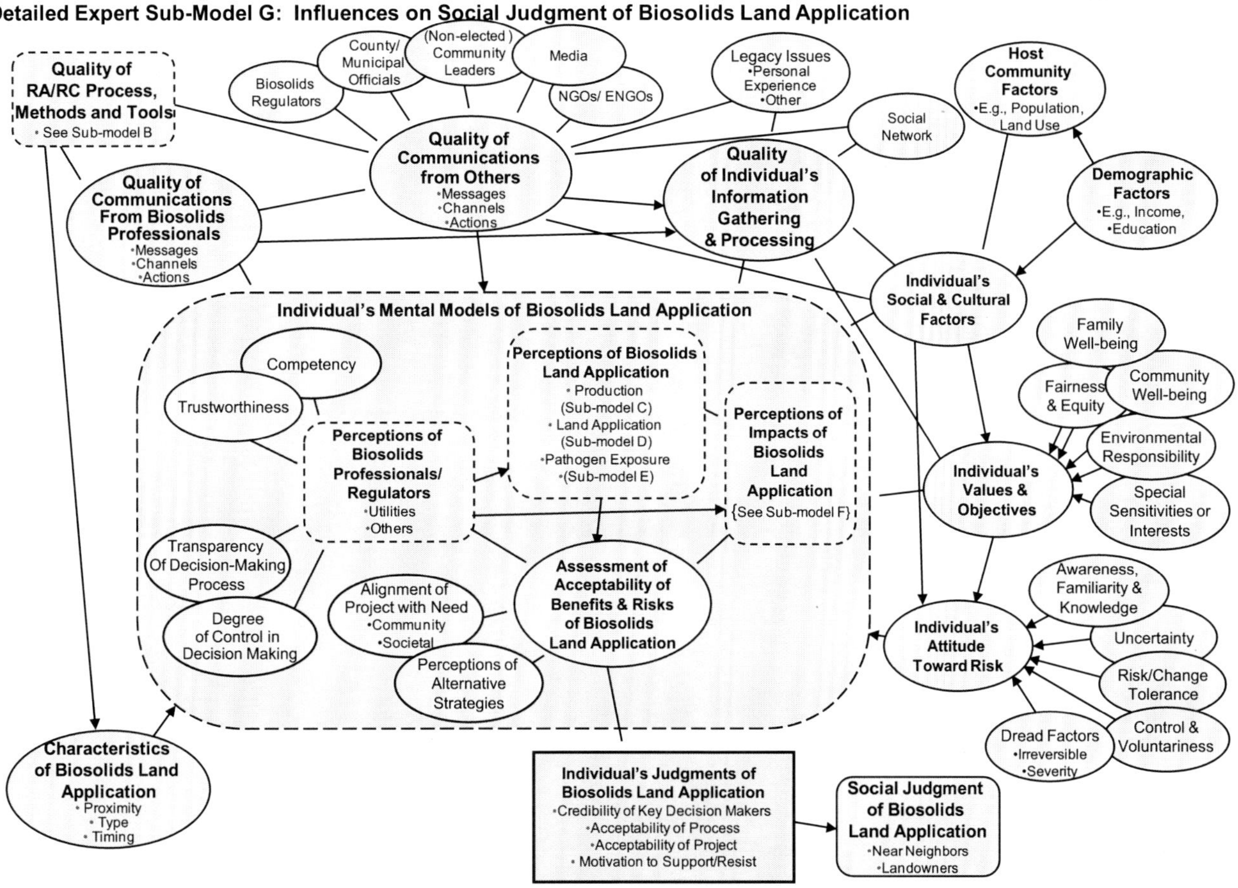

Figure A-8. Detailed Expert Sub-Model G.

The Sub-Model depicts a number of influences on *Individuals' Mental Models*, including:

- *Characteristics of Biosolids Land Application*, which include proximity, type, and timing of biosolids land application.
- *Quality of Communications from Others* which, as the Detailed Sub-Model shows, includes messages, channels, and actions from such sources as community leaders, county/municipal officials, media, NGOs/ENGOs, biosolids regulators, and the biosolids profession . The quality of these communications sources may be influenced by *Individual Social and Cultural Factors*, *Social Networks*, *Legacy Issues*, the *Quality of RA/RC Process, Methods and Tools*, and by the *Quality of Communications from Biosolids Professionals.*
- *Quality of Communications from Biosolids Professionals*, which includes messages, channels and actions. The quality of these communications may be influenced by the *Quality of RA/RC Process, Methods and Tools*. The *Quality of Communications from Biosolids Professionals* influences both the *Quality of Communications from Others* and the *Quality of Individuals' Information Gathering and Processing*.
- *Quality of Individual's Information Gathering and Processing*, which can be thought of as the quality of the information that individuals seek and use when forming judgments, and how this shapes the individual's decision making. The *Quality of Information Gathering and Processing* is directly influenced by the various sources of information that individuals actively seek and/or receive.
- *Individual's Social and Cultural Factors* which influence not only the *Individual's Mental Models* but also the *Individual's Values and Objectives*, the *Individual's Attitude Toward Risk*, the *Quality of Individual's Information Gathering and Processing* and the *Quality of Communications from Others. Individual's Social and Cultural Factors* are influenced by *Host Community Factors* and *Demographic Factors.*
- *Individual's Values and Objectives*, which are how individuals value the potential project impacts. These may include family well-being, fairness and equity, community well-being, environmental responsibility, and special sensitivities or interests. *Individual's Values and Objectives* may be influenced by *Individual's Social and Cultural Factors* and may influence not only the *Individual's Mental Models* but the *Individual's Attitude Toward Risk* and the *Quality of Individual's Information Gathering and Processing.*

Individual's Attitudes Toward Risk, which are the underlying factors by which an individual assesses various risks, including factors such as awareness, familiarity and knowledge, uncertainty, risk/change tolerance, control and voluntariness, and dread factors. These factors have been shown to have significant influence on people's judgments of the acceptability of particular risks or opportunities.

APPENDIX B

TULSA CASE STUDY APPLICATION OF THE STRATEGIC RISK COMMUNICATIONS PROCESS

B.1 Introduction

In 2008, the Water Environment Research Foundation (WERF) and its subscribers recognized the need to develop a systematic, science-based approach to building and conducting effective outreach and dialogue with key stakeholders. As part of a 2008 Biosolids Research Challenge[5], Decision Partners adapted a state-of-the-science Strategic Risk Communications Process[6], and supporting guidance tools and materials that can be used by biosolids professionals managers and other WERF subscribers to facilitate effective communications with stakeholders, thereby enabling biosolids professionals and their stakeholders to make well-informed decisions and take appropriate actions. This research, described in the main body of this report, should significantly improve WERF subscribers' ability to design and conduct effective risk communications, primarily through outreach and dialogue with their stakeholders – the people in the communities where biosolids are land-applied.

As part of the Research Challenge, Decision Partners adapted and applied the process, supported by its methods and tools, with two case study teams of biosolids professionals. One case study involved a team from the City of Tulsa, Oklahoma, and the other was a team comprising several members of the Virginia Biosolids Council. Based on the insight generated through these two case studies, the researchers developed a Primer (accompanying this report) for biosolids professionals to apply the Strategic Risk Communications Process and its methods and tools to conduct more effective dialogue and outreach with community stakeholders in the communities where they operate. This approach will be of particular interest to those responsible for managing biosolids land application and communicating with stakeholders.

This appendix describes how the Strategic Risk Communications Process were applied in this case and presents the key products, research findings and guidance developed in each step. Decision Partners worked with the Tulsa biosolids team to conduct the first four steps of the process, and to provide concrete guidance on implementing the last two steps of the process.

The case study objectives were to apply the Strategic Risk Communications Process in order to:

- Elicit the values, priorities, and informational needs of landowners who receive land application and near neighbors who live near land application sites regarding the Tulsa biosolids land application program.

---

[5] *Through the Water Environment and Research Foundation (WERF) Research Challenge SRSK2R08, Decision Partners collaborated with other researchers to develop methods and tools to assess, evaluate and communicate potential pathogen risks associated with biosolids land application. See WERF Research Challenge SRSK2R08 Final Report for more information.*

[6] *The Strategic Risk Communications Process was developed for Health Canada and the Public Health Agency of Canada (PHAC) by Decision Partners, November 2006. The Process is detailed in the Strategic Risk Communications Framework and Handbook (see www.riskcommunications.gc.ca).*

- ♦ Use the research results to support development of risk communication strategies and tools to help the City of Tulsa Biosolids Program improve their stakeholder engagement and communications related to biosolids land application.

This appendix complements this report's accompanying *Primer for Biosolids Professionals: Conducting Effective Community Outreach and Dialogue on Biosolids Land Application* (the Primer). It describes how the steps of the process were applied and presents the key products, research findings and research-based guidance developed in each step. The remainder of this report is outlined against the steps of the process. Prototypical communications plans and materials that were developed through the case study (e.g., sample meeting agendas, presentation slides and speaking notes, a biosolids pamphlet, and draft application site signage) are included as Appendix C.

B.1.1 Overview of the Strategic Risk Communications Process

Strategic Risk Communications is the purposeful process of skillful interaction supported by appropriate information to enable well-informed decision making and action. The Strategic Risk Communications Process is a step-by-step process to help biosolids professionals design and conduct effective community outreach and dialogue on biosolids land application (Figure B-1). It can be tailored to fit the unique needs of a community or a specific issue or challenge with key community stakeholders. These stakeholders include landowners, people who live near land application sites and other community officials. The process can be scaled to address a specific application, or to provide an approach for an organization that is preparing for land application in many communities. And, it supports continuous learning and continuous improvement.

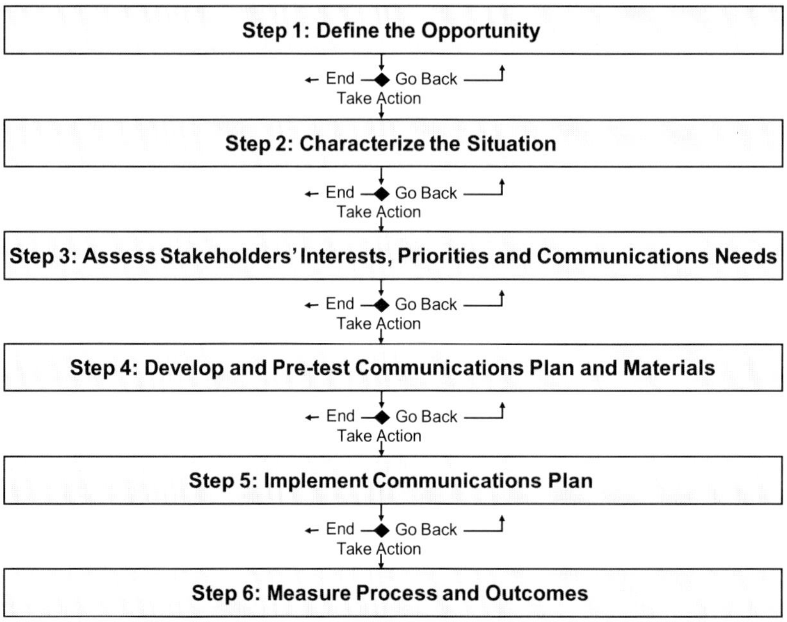

Figure B-1. Strategic Risk Communication Process for Outreach and Dialogue on Biosolids Land Application.

More detailed information on the Strategic Risk Communications Process is provided in Chapters 1.0 and 4.0 of this report and in the accompanying Primer.

B.2 Step 1: Define the Opportunity

In the first step of the process, a Biosolids Communications Team, led by the program manager, is formed. The team develops an *Opportunity Statement* – a clear statement of the current community outreach and dialogue opportunity for the biosolids program, including its goals, objectives and expected outcomes, and develops a plan for implementing the process.

B.2.1 Team Orientation

The first task of the case study was to form the Tulsa biosolids team, which included:

- Tom Krueger, Brian Bolte, and Cindy Crider of the City of Tulsa Public Works Department;
- Sara Eggers, Katherine Sousa, and Sarah Thorne of Decision Partners;
- Alan Hais of WERF; and
- Patrick Gurian of Drexel University (representing the WERF Research Challenge Biosolids Risk Assessment Team).

The members of the team provided valuable insight and supported the research that was conducted for this case study.

The case study was initiated through a two-day workshop with the Tulsa biosolids team on May 28 and 29, 2009. (See Appendix B-1 for the Orientation Workshop Agenda.) The purpose of the workshop was to:

- Orient the Tulsa biosolids team to the WERF Research Challenge and the Strategic Risk Communications Process.
- Support the team in adapting and applying the first two steps of the process (*Define the Opportunity* and *Characterize the Situation*) to meet the specific challenges and opportunities of the Tulsa biosolids team members.
- Prepare for empirical stakeholder research as part of Step 3 of the process (*Assess Stakeholders' Interests, Priorities and Communications Needs*.

The workshop was conducted at the City of Tulsa Public Works. It included a tour of the facility and the land application sites.

B.2.2 Developing an Opportunity Statement

The Tulsa biosolids team next drafted the following Opportunity Statement to describe the project scope and desired outcomes, identify project and risk communications goals and objectives, and define process deliverables and communications outcomes.

Opportunity Statement

The opportunity of the Tulsa biosolids team is to ensure the long-term sustainability of our business and the societal value it provides by systematically building key stakeholder support for, and effective management of, our current and future biosolids products and operations.

As a pilot for the WERF Biosolids Challenge, our work over the next six months will focus on applying the Strategic Risk Communications Process in order to:

- *Discover key stakeholder perceptions, interests, and priorities related to our products and operations.*
- *Proactively engage key stakeholders in dialogue about the benefits and risks of our current and future products and operations.*
- *Effectively address stakeholders' interests and priorities through dialogue supported by relevant information.*

B.3 Step 2: Characterize the Situation

Step 2 of the process was also conducted as part of the May workshop. In this step, the Tulsa biosolids team conducted a preliminary *technical assessment* to understand the necessary scope and context of the biosolids land application program in the community(ies), defining the various related technical, social, regulatory, environmental, economic, and safety considerations. The team also conducted a *preliminary stakeholder assessment* to determine the key stakeholders, and the team's current understanding of those stakeholders' interests and priorities related to biosolids land application in their community.

B.3.1 Summary of Technical Analysis

The team was facilitated in discussion on key technical, regulatory and other issues related to biosolids land application in Tulsa. The Tulsa biosolids team members also provided summary documents describing these key issues in their local communities.

The discussions in this step were guided by the **Expert Model of Influences on Integrated Risk Assessment and Risk Communications (RA/RC) System for Biosolids Land Application,** which was developed as part of the WERF Research Challenge. An expert model is a summary of relevant knowledge about a topic, prepared in a way that facilitates risk analysis and risk communications. As a formal integration of knowledge, the expert model summarizes the knowledge needed to make judgments about the topic and issues related to it. The Expert Model developed for this project provides a "systems" view of biosolids land application management, identifying the key stakeholders, processes and potential outcomes in that system. More information on the Expert Model is found in Appendix A.

B.3.2.1 Background on the Tulsa Biosolids Land Application Program

The City of Tulsa Public Works oversees the biosolids program. In 1984, an aggressive industrial pre-treatment program was implemented which enabled meeting standards for land application. Biosolids operations began in 1986. Biosolids are produced by mesophilic anaerobic digestion. (In past years, lime stabilization was used to produce biosolids, but the process was changed due to odor issues.). They are land applied in one of two ways: liquid, which is stored in a holding lagoon until land applied (Northside operations) or a 18% cake, which is a dewatered belt-filter press (Southside operations).

The team decided to focus on the cake biosolids produced by the Southside operations. An average of 17MG (3,900 dry tons) of liquid biosolids and 35,500 cubic yards (4,700 dry tons) of cake is applied annually. Biosolids from the City of Tulsa Public Works Department are applied in Tulsa, Osage and Creek Counties. Demand for biosolids land application outstrips supply and there are several area farmers and ranchers who have expressed interest to the City of Tulsa Public Works Departments about participating in the biosolids program.

The Biosolids Program is operated by one supervisor and seven full-time staff. Two contractor hauling companies are involved in the Tulsa biosolids operations and have contributed best practices for working with contractors. It does not have a specific communications program.

B.3.3 City of Tulsa Biosolids Program Stakeholder Map

At the workshop, the team also conducted a preliminary stakeholder assessment using worksheets provided by Decision Partners. The purpose of the Assessment was to identify the key stakeholders of biosolids land application in Tulsa, and develop hypotheses of their values, priorities, and information needs related to biosolids land application.

Figure B-2 shows the Stakeholder Map that was developed by the Tulsa biosolids team, identifying the stakeholders of a sustainable biosolids business in and around the City of Tulsa. For the purposes of the case study, the team identified the following stakeholders as priority stakeholders:

- Landowners: Individuals who offer their property for biosolids land application. Landowners who choose to apply biosolids on their land may receive direct benefits and may, therefore, have a different stake in projects. The Tulsa team saw landowners as biosolids communication partners, playing a potentially important role in communicating with neighbors and other community stakeholders. Also included in this cohort were potential landowners – farmers and ranchers who had expressed interest in biosolids land application but had not yet received them.
- Near Neighbors: Individuals who reside in or own property within approximately one mile of a current or potential biosolids land application site. Near neighbors and other community members may not receive direct benefits from biosolids land use (e.g., payment or soil amendments), but may otherwise be affected by land application and risk management decisions (e.g., bear a real or perceived risk and/or receive indirect benefits).

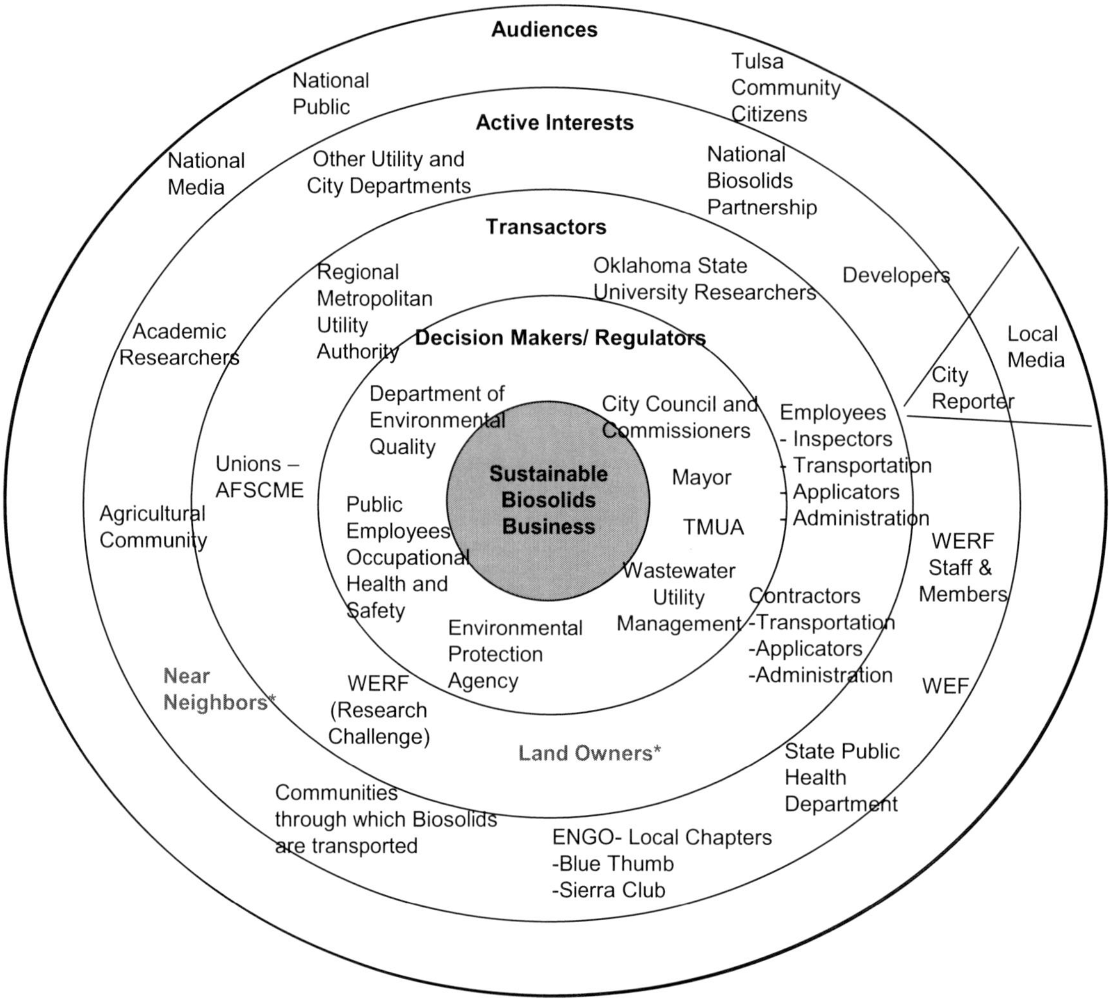

Figure B-2. City of Tulsa Biosolids Program Stakeholder Map.

B.3.4 Developing a Stakeholder Hypothesis

The final task in conducting the preliminary stakeholder assessment is to develop hypotheses of the priority stakeholders' interests, priorities, and communications needs related to biosolids land application. A stakeholder hypothesis is the team's "best guess" about the interests and priorities of the key internal and external stakeholders. It informs the preparation for and analysis of research conducted with these priority stakeholders (Step 3 of the process). At the May workshop, the biosolids team developed stakeholder hypotheses for landowners (summarized in Table B-1) and near neighbors (Table B-2).

Table B-1. Tulsa Landowner Hypotheses.

Priority Stakeholder: Landowners	
Assessment Question	Tulsa Biosolids Team Hypothesis
What they know about biosolids and biosolids management that is correct.	• They have at least a basic understanding of biosolids. • They know the agricultural and economic benefits to themselves as farmers and ranchers. The see the benefit to their property. • They know how the land application process works. • They know their neighbors don't generally like biosolids land application. We know this based on conversations with farmers. We do not have a formal process for learning what they know or what questions they have.
What don't they know or misunderstand.	• They may not take into account the risks – they don't know (at least they don't ask) about pathogens, health implications for animals. • They may not understand the regulations and requirements that the City has to follow regarding biosolids land application. They especially may not know why we have to follow these regulations. • They don't understand why biosolids sometimes cannot be applied (e.g., on fields high in phosphorus) • They don't understand how their use of commercial fertilizer affects our ability to use the field for application.
What they want to know.	• They want to know about the soil content and nutrient value – nutrients, metals, etc., before and after application. • They want to know why they can't get biosolids more often.
Who do they trust?	• OSU Extension agents. • "Us" – City of Tulsa Public Works. They have a good relationship with "us". • They have less trust in state officials or other regulators.
What they value most about biosolids land application management.	• They value first-class fertilizer at no cost. They would like as much as they can get. They would like to put it where they want, when they want. • They want fairness in who gets biosolids. • They want to minimize the inconvenience of biosolids land application, particularly the timing of what is required from them (e.g., if they are removing a crop or if they have to move cattle). • They don't want to have neighbors' complain about land application (especially odor). • They don't want anything that risks their opportunity to get biosolids.

Table B-2. Tulsa Near Neighbor Hypotheses.

Priority Stakeholder: Near Neighbors	
Assessment Question	Tulsa Biosolids Team Hypothesis
What they know about biosolids and biosolids management that is correct.	• That farmers want it – so it must be a benefit to the farmers. • A means for the City to get rid of by-products. • Lots of equipment in the field – lots of trucks. • Most/some know about farm operations. *These are primarily assumptions. Much is based on conversations with farmers. We rarely speak to neighbors directly except when they come to the fields to talk to us during a land application.*
What don't they know or misunderstand.	• What biosolids are. • They assume farmers are getting paid to take biosolids. • They think dry biosolids look like manure and think wet sludge looks like oil. • They don't consider that other forms of fertilizers may have odor and health risks associated with them. • They may not know the benefits for the whole system. • They are contributors to the system. • They may not know where their sewage goes. • They don't understand why biosolids are taken from Tulsa to other communities. • They are not aware of the regulations. • They don't know how to stay off areas where biosolids have been applied.
What might they want to know?	• What it is and what is in it? • Are there medicines/drugs in it? • Why is it here and why now? • Impacts on health, animals (including pets and wildlife). • Why there is an odor. • Why they weren't notified of the biosolids land application. • Where they can get it (many of these neighbors are also farmers/ranchers who would be interested in biosolids).
Who do they trust, and why?	• They don't trust "us". – Not a customer of Tulsa Utility. • May trust the farmer recipient. • They may trust state regulators like DEQ. • Some have called media.
What communication methods do they prefer?	• We don't know. Tulsa biosolids team hasn't had direct communications with them. Communications in past have happened through DEQ.
What do they value related to biosolids risk management?	• They value their peace of mind and way of life out in the country. • They want to be safe. • They don't want odor or other disruptions. • They probably would like input into whether the biosolids are applied nearby.

B.4 Step 3: Assessing Stakeholders' Priorities, Perceptions, and Communications Needs

From July to September, Decision Partners worked with the Tulsa biosolids team to conduct in-depth mental models research with two cohorts: 12 near neighbors and 12 landowners. This research was conducted as part of Step 3 of the Strategic Risk Communications Process.

B.4.1 Background on the Research Method

Decision Partners used an expert models/mental models research approach to conduct the stakeholder research. With its ability to characterize people's interests, perceptions, priorities, and communications needs in a structured way, this method can be used to inform the development of communications strategies and materials to address those priorities and needs. It typically involves individual, one-on-one, confidential interviews, either in person or by telephone. The method is well suited to addressing complex risk management and risk communications challenges, which require in-depth understanding of the influences on people's perceptions, judgments, and their resulting decisions and actions. More detail on the method is found in the body of the full report (SRSK2R08) and the accompanying Primer.

B.4.2 Research Preparation

The Tulsa biosolids team developed a detailed research plan in preparation for the mental models research. This included development of a sample and recruitment plan, development of an interview protocol and analysis and reporting plan. More details on the research preparation and analysis are included in Chapter 3.0 of this report. The key features of the research are described below.

The team supported the development of the research sample which included:

- 12 "Landowners": Individuals who own property on which biosolids have been applied in Tulsa, Orange or Creek Counties; and people who have expressed interest in biosolids land application to the City of Tulsa Public Works Department. The City of Tulsa provided a list of landowners.
- 12 "Near Neighbors": Individuals who live within approximately one to three miles of a biosolids land application site; and other people who live in small communities near biosolids land application sites. Neighbors were identified as living within a set distance from the land application sites, using Google Maps® and Yellowpages.com®.

A mental models interview protocol was developed to solicit near neighbors' and landowners' thinking on topics related to biosolids land application, including their values, priorities, perceptions of the benefits and risks associated with biosolids land application, with particular attention paid to pathogens.

While tailored to each of the two cohorts, the protocol generally followed this outline:

- General thoughts on biosolids, including, why and how biosolids may be land applied, and thoughts after hearing brief descriptions of biosolids.
- Benefits and risks associated with biosolids land application, including potential pathogens in biosolids land application and their impacts to land, water, animals, and people. A description was provided.

- Thoughts on the decision-making process for biosolids land application, including how decisions regarding biosolids land application are made, who is involved, and how they are responsible.
- Communications related to biosolids land application, including most important information in communities where biosolids are applied, and thoughts on most effective ways for biosolids managers to communicate about biosolids.

B.4.3 Conducting Research Interviews

Decision Partners researchers conducted 24 phone interviews between July 9 and September 21, 2009. Table B-3 and Figures B-3 and B-4 summarize the key demographic characteristics of the research sample. Interviews averaged 37 minutes in length, ranging from 24 to 53 minutes. Interviewees were offered an honorarium of $30 for their participation. The mental models interviews were audio-recorded with the permission of the participants and transcribed.

Table B-3. Summary of Tulsa Interviewee Demographic Characteristics.

	Near Neighbors	Landowners
Number of Interviewees	12	12
Interview Dates	July 9, 2009-September 21, 2009	July 10, 2009-August 7, 2009
Interview Lengths	Range: 24-40 min. Average: 33 min.	Range: 30-53 min. Average: 41 min.
Number (%) Females	3 (25%)	2 (17%)
Number (%) Children Under 18	6 (50%)	5 (42%)

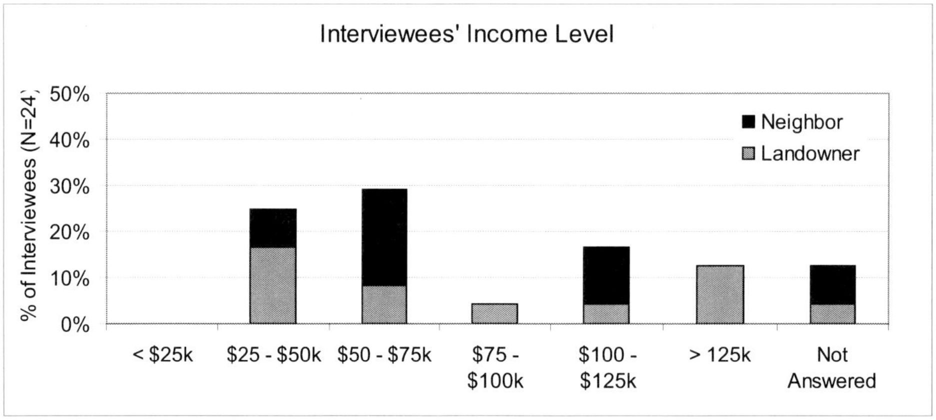

Figure B-3. Tulsa Interviewees' Income Level.

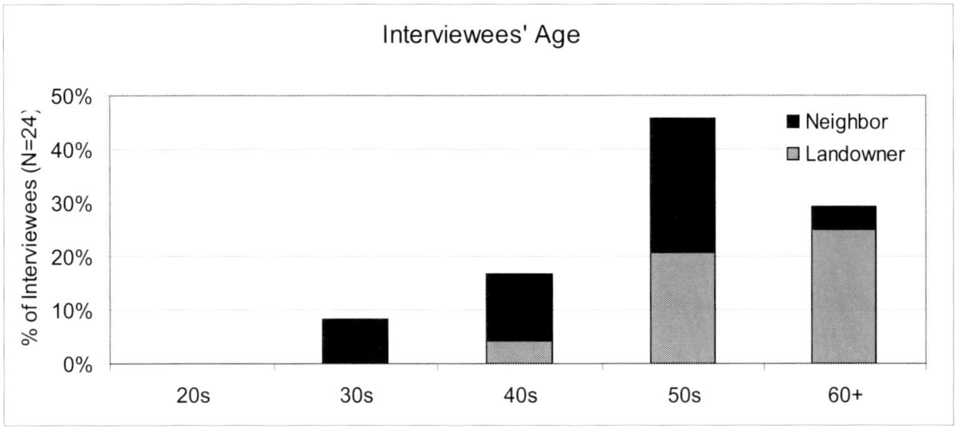

Figure B-4. Tulsa Interviewee' Ages.

B.4.4 Key Research Results

Decision Partners prepared and presented the research report to the Tulsa biosolids team on November 4, 2009. The following are the key findings from the mental model research conducted with landowners and neighbors as part of the Tulsa Case Study. A more detailed summary of research findings, compared against the Tulsa team's Stakeholder Hypothesis, are presented in Appendix B-2.

- Interviewees were generally familiar with biosolids and aware that biosolids are land applied in their community.
- A variety of mental models regarding biosolids and biosolids land application emerged. Some had specific knowledge of biosolids treatment; however, it appears that many interviewees have very little understanding of treatment processes.
- Landowners appeared to be very familiar with pre- and post-application requirements. Nearly all neighbors spontaneously expressed concern about whether the regulations were being – or could be – followed, citing primarily lack of compliance and lack of effectiveness.
- Landowners cited direct positive agricultural and economic impacts of biosolids land application. neighbors cited more indirect positive impacts, including wastewater management and agricultural benefits to farmers and ranchers.
- Almost all interviewees attribute odor to biosolids land application. Neighbors believed this was a significant impact and attributed health and quality of life impacts to odor. Landowners believed it was less significant, because it is "minor", "short-lived" and "worth it".
- Most interviewees believed that biosolids land application poses a risk of pathogen exposure. Many also generally believed that pathogens occur "naturally" in many agricultural sources. Landowners were significantly less concerned about pathogens than neighbors were, citing their trust in the utility and their belief that the biosolids treatment processes reduce the amount of pathogens to "safe" levels.

- Other negative impacts that interviewees cited included hazardous waste, runoff, over-application, contamination, animal health, and long-term uncertainty.
- Landowners expressed confidence in the safety of biosolids land application, citing their trust in the regulations and the City of Tulsa and the "naturalness" of biosolids. Neighbors expressed much less confidence in the safety of biosolids land application, citing their lack of trust in decision makers, lack of confidence in the regulations and the uncertainty of biosolids.
- Interviewees believed that the most important considerations in decisions regarding biosolids land application are the quality and oversight of regulations, the safety of biosolids and the impact on neighbors.
- Nearly all landowners expressed trust in biosolids decision makers generally speaking and most expressed trust in the utility. Most neighbors spontaneously mentioned that they do not trust the government and/or city officials. Some neighbors mentioned that they do not trust the utility, citing motivation, regulations and site selection as the key reasons for their lack of trust.
- Near the end of the interview, interviewees were asked to rate themselves as generally "in favor", "not in favor" or "undecided" regarding biosolids land application in their community. All landowners were in favor of biosolids land application, citing improved growth of grass and hay, economic benefits, and the ability to recycle nutrients from waste. Three neighbors rated themselves as generally "in favor" of biosolids land application. Nine neighbors, however, said they were "not in favor", citing the odor, health and quality of life impacts, and uncertainty.
- Interviewees commented on several opportunities to improve the quality of engagement with people in the community about biosolids, including having more transparency in decision making, communicating in advance and providing information on the benefits and risks.

B.4.5 Research-Based Recommendations for the Tulsa Case Study Team

Using the insight generated from the first three steps of the process, including the research results from the neighbors and landowners, Decision Partners prepared recommendations focused on improving Tulsa's outreach and dialogue with community stakeholders in the communities where the Tulsa Biosolids Program is operating.

These recommendations were developed to focus Tulsa's efforts – and future initiatives – on the highest-value opportunities described by the Opportunity Statement. They included guidance on how to effectively engage local community members in dialogue; foster effective coordination, communication, and collaboration among key partners; and develop effective and consistent communications materials to support community dialogue.

Based on the research results, Decision Partners offered the following strategic recommendations to the Tulsa biosolids team:

1. Continue to frame biosolids land application as a beneficial *product* that when used appropriately can yield benefits to people in this community and more broadly.

2. Look for opportunities to engage local community members in dialogue about wastewater treatment and the opportunity biosolids presents for beneficial use. A specific recommendation was to conduct small dialogue meetings in advance of the biosolids land application projects. The Tulsa biosolids team should also work to support landowners' engagement with their neighbors and other community members. In addition, the website can be enhanced to enable dialogue on biosolids with community members.

3. Use the research findings to develop core communications to support dialogue. Such communications materials should:

   - Enable learning about wastewater treatment generally and biosolids treatment specifically.
   - Address landowners' and neighbors' concern about the safety of biosolids – and how they can be assured that they are safe.
   - Facilitate a shared understanding among neighbors and landowners of the decision-making process.
   - Address neighbors' concern about odor.
   - Clearly communicate about specific projects – where and when.
   - Clearly communicate to landowners biosolids land application restrictions, requirements, and safe practices, including their underlying rationale.

4. Develop standard communications materials, including:

   - A dialogue presentation, which can serve as a basis for neighbor engagement meetings and other dialogue opportunities. This presentation should provide a basic discussion on the City of Tulsa biosolids land application program and address the key topics stakeholders identified in the research.
   - A brochure on the City of Tulsa biosolids land application program, which could be adapted as the core text for website. This brochure should be aligned with the dialogue presentation material.

5. Develop a plan for developing and implementing signage at current application sites. The team hypothesized that signage would be appreciated by neighbors. In addition, a proactive step in developing and using signage could be an important initiative demonstrating Tulsa as a leader in the community.

6. Pre-test materials with landowners and neighbors to ensure they are interpreted as intended.

7. Develop effective communication for landowners, as partners in the biosolids land application program. Such communications may include clear guidelines for communicating with their neighbors. It also includes clear guidelines on landowners' role and responsibilities as partners in the land application program. *Note*: Decision Partners provided some specific guidance for the City of Tulsa Public Works, not included in this appendix.

Preliminary recommendations were reviewed by the Tulsa biosolids team during a team meeting on November 4, 2009. The recommendations were refined in light of this input and a plan was developed to design and pre-test a draft plan and supporting materials.

B.5 Step 4: Develop and Pre-Test Communications Plan and Materials

In Step 4 of the process, the Biosolids Communications Team develops a Communications Plan for community outreach and dialogue that addresses what the team learned about people's priorities, perceptions, and communications needs.

B.5.1 Development of Draft Neighbor Engagement Plan

From November 2009 to February 2010, the biosolids team worked to develop a draft plan for conducting neighborhood dialogue meetings. This plan included the process for communicating with and engaging neighbors in dialogue about upcoming biosolids land application. Table B-4 below summarizes this process, showing the actions to be taken in preparation for the meeting, by whom and by when.

Table B-4. City of Tulsa Biosolids Program Neighbor Engagement Plan.

DATE	ACTION	WHO
Four weeks before application	City staff meets with the landowner to confirm application and discuss the dialogue session and biosolids signage.	City Staff Landowner
	OPTIONAL: landowner posts "No Trespassing" signs on his or her property.	Landowner
Four weeks before application	Plan for biosolids dialogue session: • Determine date, time and location for the session. • Confirm that landowner will attend the session. • Reserve location for session. • Develop contact list of neighbors who live approximately one mile from the application site.	City Staff Landowner can support.
Three weeks before application	Send out Biosolids Pamphlet to landowner and neighbors living within approximately one mile of the application site. Include an invitation to attend the dialogue session • Neighbor will be asked to call the City Staff to RSVP to the event at least two days before the session.	City Staff
Two weeks before application.	Develop the agenda for the session. • Line up City participants. • Determine if any other City departments need to be contacted in case they get calls. • Plan for refreshments for participants. • Make sure necessary equipment is available (e.g., projector). • If possible, call all invitees to confirm their participation 1-2 days before. • Review neighbor Engagement Dialogue Presentation. Make any needed modifications and practice dialogue. • Prepare meeting agenda. (The Pre-test Agenda in Appendix B-3 provides an example.)	City Staff
At least one week before application	Host a 1.5 hour neighbor dialogue session	City Staff
Within one week following neighbor dialogue session	Follow up with any neighbors or other dialogue session participants as needed to address questions or respond to specific requests for information.	City Staff
Three days before application – after dialogue session.	Post biosolids application signage indicating the field is a biosolids land application site (landowner must allow this).	City Staff
Thirty days after application (minimum)	Remove land application sign.	City Staff

B.5.2 Development of Draft Communications Materials

From November 2009-February 2010, the Tulsa biosolids team also worked in collaboration (supported by a graphic designer) to develop the following communications materials:

1. Biosolids pamphlet: A basic pamphlet on biosolids was developed to provide general information on Tulsa's biosolids program and address the topics identified in the research. The pamphlet is aligned with the presentation and speaking notes, using a similar question and answer format. It includes pictures of the Tulsa biosolids team, the application process and the "before" and "after" demonstration of the beneficial use of biosolids on local land. The final dialogue presentation is included in Appendix C-1.

   This pamphlet was developed with several potential uses in mind. It can be distributed to landowners at meetings, through the mail or door-to-door. It can also be distributed through more general outreach opportunities. Finally, the content of the brochure can be easily adapted as text on the website. By so doing, the Tulsa team will also have the opportunity to add further detail about the biosolids program for people who want the additional information (e.g., links to the Oklahoma Department of Environmental Quality website).

2. Generic presentation and speaking notes: A presentation was developed to support communications about biosolids land application with neighbors and other key community stakeholders. The presentation was intended to be about 20 minutes in length. It was written in a question and answer format, focusing on the key topics that were identified in the research. It was also written in the voice of the Tulsa biosolids team, focusing the discussion on the local context. Although written with the landowners and neighbors in mind, the presentation can serve as a template for developing presentation materials to engage other key stakeholders in dialogue, such as City Council members and the local media. The final dialogue presentation is included in Appendix C-2.

3. Prototypical signage for the City of Tulsa Public Works to post before, during and after a biosolids land application. (See Appendix C-3.) The signage was designed to focus on biosolids as a *product* with beneficial use. It provided an approximate date of the land application and contact information for the City of Tulsa staff.

B.5.3 Conducting Pre-tests of Communication Plan and Materials

Pre-testing messages and information materials is critical to ensure the stakeholders will understand them and that their messages have the intended effect. Decision Partners facilitated the Tulsa biosolids team in conducting pre-tests in two small workshops (one with landowners and one with neighbors) on March 9, 2010. The purpose of the pre-test was to test the neighbor engagement plan (Table B-4) and test the supporting dialogue presentation, dialogue brochure and site signage. The pre-test also provided the team members from the City of Tulsa with practice and coaching on dialogue skills.

The biosolids team carefully prepared for the pre-tests in February and March 2010. The interviewees who participated in the mental models research and their adult family members were invited to attend. Six landowners chose to participate in the pre-test and four neighbors chose to participate. The pre-tests were held in the public library in one of the small communities around which biosolids are applied. (The meeting agenda is found in Appendix B-3.)

The meeting began with introductions and an overview of the meeting purpose and agenda. Brian Bolte from the City of Tulsa next delivered the biosolids engagement presentation. Following the presentation the biosolids team members conducted individual interviews with participants, using the protocol found in Appendix B-4. Following the individual interviews, participants were asked to read the biosolids pamphlet. After allowing time for the participants to read the brochure, the team facilitated a group discussion on all of the overall engagement session and the communications materials.

The following is a summary of pre-test results from the landowner dialogue session (N=6):

- The landowners all appeared to be very appreciative of the chance to talk to the City staff about the biosolids program. They were very engaging in their discussions and their suggestions.
- The landowners believed that the presentation was "complete", "informative" and "understandable".
- They said that they agreed that biosolids are a product which has beneficial use.
- They believed that the presentation covered the important benefits and appreciated that those benefits were demonstrated in the example photos. They said they took away from the discussion the fuller range of benefits beyond strictly agricultural (which they know first-hand) and that that land application keeps such a "beneficial product" out of the landfill. They said that they appreciated learning more about the treatment and testing processes.
- They said that the information on safety did not raise any questions in their minds, but pointed out that their neighbors may have more questions.
- They raised questions about odor and the chemicals used to treat biosolids, since both of these topics are a big concern to their neighbors. They also indicated they wanted more information on the regulations, particularly *why* they are in place.
- They suggested that more information comparing biosolids to other fertilizers would be helpful, as would information on how decisions are made about where and when to apply biosolids.
- They believed that meetings with neighbors would be very beneficial, and suggested that landowners and perhaps a regulatory or USDA agricultural extension official could participate in the meetings.
- They believed that signage would be beneficial and that neighbors would appreciate being notified of the application before it occurs.

The neighbors who participated in the dialogue session were generally critical of the nearby biosolids land application (as interviewees in the research, they rated themselves "not in favor"). Although it does not appear that the dialogue session "changed their minds", it does appear that, the neighbor participants all appeared to appreciate having the chance to engage in conversation with the City staff. The following is a summary of the pre-test results from neighbor dialogue session (N=4):

- The neighbors generally found the presentation to be "informative", "a good overview" and "well put together". One neighbor felt it was "too basic". Another neighbor said she at first believed the purpose of the meeting was to try to "sell something".
- The neighbors indicated that they appreciated that the staff from the City of Tulsa was trying to better communicate with them.
- Odor was a key discussion topic. Neighbors were critical of the communication on odor – they believed that the odor from biosolids is "offensive". They also believed that the discussions on odor were "downplayed". As one said: "It doesn't smell like compost. It smells like sewer." [The brochure originally stated that the biosolids produced by the City smelled like high-quality compost.] One said it severely affected his/her quality of life, and invited the team to come "have coffee with me after you apply".
- The neighbors offered varied comments on biosolids safety. A few said it was their number one concern. They questioned whether the "pathogens can become airborne, since there is an odor". One commented on seeing runoff from fields go near houses. Another, however, said s/he was comfortable with the safety because it is well-managed.
- All neighbors said they wanted more information on the testing and regulations, particularly *why* the biosolids land application regulations are in place. One asked: "What makes biosolids safe after 30 days?"
- A few neighbors said that they wanted more concrete, "unbiased", and "certified" data demonstrating the safety of biosolids.
- A few neighbors raised concerns about the truck traffic on the small county roads. They said that the roads are being damaged and are not repaired. They asked whether the City could help repair the roads.
- It appeared that the presentation did not "change the minds" of the neighbors who had rated themselves "not in favor" of biosolids land application in the research. As one said: "The presentation is not going to change the product."
- Other discussion topics that were raised included whether community members have any say in the decisions or any recourse; weighing the societal benefits of managing a "problem" with the risks to individuals who live near the land application: "What is fair?" Participants mentioned that they have septic systems. It was discussed whether septic systems are included in the biosolids program (often, yes).
- Neighbors believed there is benefit to having meetings with neighbors about biosolids. The appreciated the chance to voice their concerns. However, they said that they did not expect that the dialogue would bring about any real change.

B.5.4 Refinement of the Communication Plan and Materials

Using the insight generated from the pre-tests, the team revised the communications materials. Some changes involved wording modifications to clarify text that had raised questions. Some changes involved the addition or deletion of topics that were identified as being more or less important. Key changes to the biosolids materials were as follows:

- Clarified the definition of biosolids, where they come from, and how they are treated to address some confusion raised in the pre-test. Septic systems were included as a source of wastewater.
- Clarified the different types of benefits: agricultural, environmental, and community benefits.
- Included an explanation of the 30-day land restriction in the presentation and the brochure: "As an added safety precaution, the land is not to be harvested, grazed by livestock or used by people for at least 30 days after the application. This is to allow for any possible remaining pathogens to die off naturally."
- Removed a slide from the presentation that communicated what people could do *themselves* to ensure their safety with biosolids and other potential sources of risk on a farm ("What you can do to assure safety"), such as staying out of fields, wearing protective clothing and washing produce. This discussion caused confusion and was determined by the group to be not important to the conversation.
- Removed the statement from the presentation and the brochure stating that biosolids smells like high-quality compost. This raised criticism from the neighbors.
- Simplified the language on the sign to make the text easier to read. The results of the pre-test suggest that a simple sign is best.
- Added detail to the speaking notes about possible questions that could be raised about environmental factors that may influence biosolids odor, and how they may best be addressed.

Based on the pre-test results, Decision Partners also recommended the Tulsa biosolids team develop of a set of more detailed or technical questions and answers, which can help the Tulsa team anticipate questions of such nature that may arise in discussion. Example questions included: *What chemicals are used in the treatment process?* and *Why is there a 30-day restriction on access to a field where biosolids have been applied?* Based on the results of the pre-test, Decision Partners drafted (Appendix C-3) a set of questions and provided some guidance on possible sample responses and other references to help the Tulsa team address these questions.

B.6 Next Steps and Future Opportunities

With the plan and supporting material developed, tested, and refined, the Tulsa biosolids team is now ready to begin to implement the plan and distribute the materials, beginning in 2011. Continual refinement to the plan and the materials will be necessary, as the team continues to learn more through their engagement with their landowners and neighbors. The plan may also be extended to include the landowners who receive the liquid biosolids produced by the Northside treatment facility (and their neighbors) and broader engagement with City of Tulsa community officials, for example.

The Tulsa biosolids team saw great value in working through the Strategic Risk Communications Process. They believed that the need for effective engagement with key stakeholders that addresses their interests and priorities is clear. The team said they learned that it is a "risky assumption to assume that everything is fine", and that they would "rather be proactive than end up with a crisis down the road." They felt this was true not just for the City of Tulsa's program, but more broadly for the biosolids community.

The team saw the value, through application of the process in identifying the full range of stakeholders, focusing on key stakeholders and better understanding their interests, priorities and information needs. They saw specific value in the Expert Model (described in the Research Challenge Final Report), which they believed provides a comprehensive "systems view" of their operations. They believed the Expert Model could be adapted to fit their specific needs and serve as a decision and communications tool, for example, by helping to track a "game plan" to mitigate, resolve, or explain issues, and avoid "crisis mode".

Finally, the team believed this project can be beneficial in the long run to help continue to move other opportunities forward (e.g., adoption of new technologies), and reduce the chance of "backsliding" as city officials and supporters change or move on. The process and the foundational work in this first cast study can be adapted for other operational issues in the future (e.g., movement to Class A, possible move of Biosolids Operations).

APPENDIX B-1

STRATEGIC RISK COMMUNICATIONS WORKSHOP WITH THE TULSA BIOSOLIDS TEAM

May 28 and 29, 2009

Location: Southside Wastewater Treatment Plant, 5300 S. Elwood Ave., Tulsa

Sponsors: Alan Hais, WERF (participating by phone when possible) and Tom Krueger, City of Tulsa.

Facilitators: Sarah Thorne and Sara Eggers – Decision Partners

Participants from the Tulsa Biosolids Team:
Tom Krueger
Brian Bolte
Cindy Crider

Introduction: As an integral part of an effective risk assessment process, Strategic Risk Communications is a purposeful process of skilful interaction supported by appropriate information to enable well-informed decision making and action on risks. The process applies a suite of methods and tools to enable effective communications with stakeholders. Its purpose is to enable both stakeholders and decision makers to make well-informed decisions and take appropriate actions on risk.

Through the application of the Strategic Risk Communications process to this WERF Research Challenge: Biosolids Risk Assessment and Risk Communications, Decision Partners will demonstrate how the process, methods and tools can be applied by a range of users responsible for land application of biosolids, including wastewater utilities, land appliers, and other government agencies. A key goal of this project is to enable WERF subscribers and other biosolids professionals to apply the process and, by so doing, build capacity for conducting effective risk communications within their own organizations. The City of Tulsa has volunteered to be the first "case study" application for this process. Working with the Tulsa biosolids team, we will adapt and apply the Strategic Risk Communications Process to the biosolids opportunities and challenges in Tulsa and surrounding communities and make recommendations for improvement.

Workshop Purpose: The purpose of the workshop is to orient the Tulsa team to the WERF Research Challenge and the Strategic Risk Communications Process and support the team in adapting and applying the process to the specific challenges and opportunities of the Tulsa biosolids operations. Participants will be coached through the initial planning steps of the process and will support Decision Partners in the implementation of the stakeholder research in the next step.

The essential Strategic Risk Communications principles, process, and tools will be discussed. Methods and tools – and the rationale for their use, will be reviewed and applied to the opportunities and challenges faced by the Tulsa biosolids team. Hands-on training with coaching on the application of the methods and tools will be emphasized.

Workshop Objectives: The objectives of the workshop are to:

- Orient the Decision Partners team to the Tulsa utility's biosolids application operations and related issues, challenges, and community outreach and communications initiatives.

- Orient the Tulsa biosolids team to the WERF Research Challenge and Opportunity and Integrated Risk Assessment/Risk Communications Approach.

- Provide the team with an in-depth understanding of the principles, process, and tools for conducting Strategic Risk Communications.

- Provide the team with hands-on coaching as they begin implementation of the first three steps of the process, customized for Tulsa:
  - Develop an *Opportunity Statement* (Step 1: Define the Opportunity)
  - Develop the *Stakeholder Hypothesis* (Step 2: Characterize the Situation)
  - Customize the *Expert Model* (Step 2: Characterize the Situation)
  - Finalize *Mental Models* Research Plan (Step 3: Assess Stakeholder Perceptions of the Risks, Benefits and Tradeoffs)

- Discuss Next Steps.

Meeting Agenda
Thursday May 28, 2009

Time	Topic	Who's Responsible
8:15 a.m.	**Meet at the Hilton for Breakfast**	All
9:00 a.m.	**Orient DP to Tulsa Biosolids Operations** • Overview of the operations • Conduct a tour of the area where biosolids are applied and through which biosolids are transported • Visit a biosolids site and see an application underway	Brian Bolte
12:00 p.m.	**Lunch with the Team (TBD)**	

12:45 p.m.	**Introduction** • Overview of the Workshop Objectives and Agenda • Overview of Research Challenge • The importance of the Tulsa case study to the WERF Research Challenge	Sarah Thorne *Alan Hais to join by phone for the afternoon?

1:00 p.m.	**Orient the Tulsa Biosolids Team on the Approach**
	• Overview of the Integrated Risk Assessment/Risk Communications Approach
	• Overview of Strategic Risk Communications Process
	    • Discuss the principles of strategic risk communications
	    • Provide an overview of the process
	    • Discuss progress to date on the WERF Challenge
		Sarah Thorne & Sara Eggers

2:00	**Break**	
2:15	**Implementing the Approach: Step 1: Develop the Opportunity Statement** • Discuss the primary activities and review the key tools in this step • Discuss Tulsa's biosolids business goals and objectives • Discuss Tulsa's stakeholder engagement goals and objectives • Discuss critical challenges and opportunities • Draft the Opportunity Statement	All, Sarah Thorne facilitate
3:00 p.m.	**Implementing the Approach: Step 2: Characterize the Situation – Conduct a Stakeholder Hypothesis** • Discuss the primary activities and review the key tools in this step • Conduct the Stakeholder Hypothesis	All, Sarah Thorne facilitate
4:15 p.m.	**Day One Wrap Up and Next Steps** • Review what we have accomplished • Discuss agenda for Day Two • Feedback from Tulsa biosolids team	Sarah Thorne
4:30	**Workshop Close**	

Agenda Friday, May 29, 2009

Time	Topic	Who's Responsible
8:30 a.m.	**Review Day 2 Agenda**	Sarah Thorne *Alan Hais to join by phone?
8:45 a.m.	**Step 2: Implementing the Approach: Characterize the Situation – Complete Stakeholder Hypothesis** (if required)	All, Sarah Thorne facilitate
9:30 a.m.	**Step 2: Implementing the Approach: Characterize the Situation – Expert Model** • Present background on Expert Modeling • Review the Draft **Expert Model of the Influences on an Integrated Risk Management and Risk Communications System for Appropriate Biosolids Land Use**	Sara Eggers
10:30 p.m.	**Break**	
10:45 p.m.	**Step 2: Implementing the Approach: Characterize the Situation – Expert Model continues** • Refine and adapt the model for Tulsa • Revisit and refine the Opportunity Statement as necessary	All, Sarah Thorne facilitate
11:45 p.m.	**Lunch**	

12:30 p.m.	**Step 3: Implementing the Approach: Assess Stakeholder Perceptions of the Risks, Benefits, and Tradeoffs**
	• Discuss the primary activities and review the key tools in this step
	• Discuss methods for assessing stakeholder perceptions
	• Review and discuss the draft mental models research plan including:
	    • Research Method
	    • Sampling and Recruitment Plan
		Sarah Thorne
		
		All, Sara Eggers facilitate

	• Interview Protocol Outline • Workplan	
1:45 p.m.	**Workshop Wrap Up and Next Steps** • Review what we have accomplished • Discuss next steps • Feedback from Tulsa biosolids team	Sarah Thorne facilitate
2:00	**Workshop Close**	

APPENDIX B-2

SUMMARY OF KEY RESEARCH RESULTS

This following is a summary of the gaps and alignments of the aggregated mental models of the 24 interviewees as part of the research that was conducted in the Tulsa Case Study.

Guiding Questions	Summary based on the research with Landowners	Summary based on the research with Neighbors

What do they know now about the topic that is correct?
 	
➢ Key findings not anticipated in the Stakeholder Hypothesis.
	• Biosolids are treated by-products (including human waste) of wastewater that can be applied to land.
	  • Most demonstrated general and some demonstrated more specific knowledge about biosolids treatment processes.
	• Biosolids produced by the City of Tulsa are currently land applied in their County.
	  • Most demonstrated specific knowledge, from direct experience, of the biosolids land application process.
	• Land application has certain restrictions and requirements before and after application.
	  • Variance: A few appear to be less familiar (i.e., demonstrate lack of compliance) with site restrictions.
	• Biosolids land application yields agricultural benefits that translate into economic benefits to landowners:
	  • Landowners see direct benefits, including richer soil, faster growth of cattle and increased crop production.
	  • Landowners see direct economic benefits, including cost savings on fertilizer and increased revenue.
	  • Potential landowners also believe in the agricultural and economic value and want biosolids land application for this reason.
	• Biosolids land application is low-cost and an environmentally sustainable way to manage ("get rid" of) wastewater by-product and can benefit the community as well.
	• The possible negative impacts of biosolids land application include odor, over-application, contamination, and land damage.
		➢ Biosolids are by-products (including human waste) of wastewater that can be applied to land.
		  • Neighbors commented on the site of the wastewater treatment plant on the Arkansas River.
		  • Most were familiar with the term biosolids.
		  • Variance: Although many neighbors were generally aware that biosolids are treated, some appeared to be less aware.
		➢ Biosolids produced by the City of Tulsa are currently land applied in their County.
		  • Most commented on direct experiences seeing (and/or smelling) land application.
		• The importance of certain restrictions on land application and requirements for landowners before and after.
		  • They express concern about the utility's and landowners' compliance with these requirements.
		• Landowners want biosolids, so it must have agricultural benefits.
		  • Most believe biosolids are low-cost or free and have some benefits as a fertilizer.
		• Biosolids land application is beneficial to the City of Tulsa because it is a low-cost way to manage ("get rid of") wastewater by-products.
		• The possible negative impacts of biosolids land application include odor, over-application and contamination.
		  • A few neighbors who were specifically asked if there had been any improvement in the level of odor commented that they had not noticed any improvement.
		• Pathogens naturally occur in many

Guiding Questions	**Summary based on the research with <u>Landowners</u>**	**Summary based on the research with <u>Neighbors</u>**
What do they know, *continued*	• One landowner mentioned an improvement in the level of odor. • Pathogens naturally occur in many agricultural sources. Biosolids land application poses small but real risk of pathogen exposure. Biosolids treatment processes reduce the amount of pathogens to "safe" levels. • The options to use sources of fertilizers, and the downsides (cost and environmental impact) of those alternatives. • Neighbors in their area do not like biosolids land application, particularly because of odor.	agricultural sources. Biosolids land application risk of pathogen exposure. • They do not receive communications (e.g., advanced notification) from the City of Tulsa regarding biosolids land application.

What don't they know or mis-understand that is consequential?	• Wastewater treatment processes.
	• The decision-making process for biosolids land application, particularly selection criteria and availability of product.
	• Risks of not complying with biosolids restrictions and requirements.
	• Landowners may be overconfident in the "safety" of biosolids treatment and land application.
	• Other specific risks associated with biosolids land application such as trace organic compounds, heavy metals, etc.
		• Wastewater and biosolids treatment processes.
		  • The role of biosolids treatment in reducing pathogens to "safe" levels. Neighbors have less confidence in the "safety" of biosolids treatment and land application.
		• Specific characteristics of pathogen risk (e.g., risk to cow's milk.)
		• The decision-making process for biosolids land application, particularly selection criteria and availability of product.
		  • Where and when biosolids can and cannot be applied.
		  • Some neighbors think that farmers are getting paid to take biosolids.
		• The use sources of fertilizers, and the downsides (cost, odor and environmental impact) of those alternatives.
		• The potential system-wide benefits of biosolids land application.
		• Other specific risks associated with biosolids land application such as trace organic compounds, heavy metals, etc.

What might they want to know?
 	
(Listed in no particular order.)
	• What is the specific value?
	  • How does the nutrient compare to other fertilizers?
	  • What is the soil content – nutrients, metals, etc., before and after application
	  • Will there be any effect on water sources?
	• Is it safe and how do we know?
	  • What are the risks? Should we be concerned?
	  • What pathogens are killed?
		• What is it and what is in it?
		  • How is it going to affect their health?
		  • Is it going to smell? Why?
		  • What are the long-term impacts?
		• Is it safe and how do we know?
		  • What is known about the safety of biosolids?
		  • Are they targeting the right pathogens?
		  • Does the treatment reduce pathogens?
		  • Are City officials doing soil samples? Will they provide records on the Internet?
		• The decision-making process:

Guiding Questions	Summary based on the research with Landowners	Summary based on the research with Neighbors
	• How do the treatment processes reduce pathogens? • The decision-making process: • How to get/use biosolids • Why they can't get biosolids more often? • Why or how their site was (or was not) selected? • Who decides where to put it and why? • What impact does the city have on it and what expenses are involved? • What are the alternatives?	• Why is it here and why now? • Why weren't they notified? • Do we have a choice? • Who puts it into the ground? • When is City going to quit doing it? • Who is responsible? Who are the key decision makers? • How do they regulate where it's placed? • What are the alternatives and are they being pursued?
Who/What do they trust, and why? (Listed in no particular order.)	• City of Tulsa Public Works, because, "They know how to treat and process", they "do what they are supposed to do to make sure the biosolids are safe", and they "follow the regulations." • Biosolids, because: • They are "natural". • "Somebody has figured out a way to make it safe." • Themselves.	• Note: Neighbors commented that they do not trust the City of Tulsa. • Landowner because "they would be in on it, he knows." Regular person who "lives out where they can relate to it." • Oklahoma State University Extension because they have "a great agricultural department." • Other trusted people in the community, such as their doctor. • Themselves.
What communication methods do they prefer? (Listed in no particular order.)	• Mailings or information included in the Utility bill. • Television/News Media • One-on-one meetings. • Internet or websites. • Town meetings. • Tours of properties where biosolids have been applied hosted by the landowners.	• Mailings or information included in the Utility bill. • Television/News Media. • One-on-one meetings. • Internet or websites. • Town meetings. • Signage.
What are the key factors that they consider when making judgments and decisions related to biosolids land application? (Listed in no particular order.)	• Agricultural and direct economic benefits to landowner. • Community benefits. • Trust in key decision makers – City of Tulsa. • Perceived fairness (e.g., site selection). • Regulations and restrictions. • Information (including soil analysis, nutrient analysis) from City of Tulsa Public Works Department. • Regulations. • Safety. • Benefits including agricultural and economic. • Health impacts. • Environmental impacts. • Impacts on neighbors (odor). • Legacy Issues (e.g., turkey manure).	• Impacts on quality of life – odor. • Health and safety impacts. • Confidence in the regulations. • Trust in key decision makers. • Uncertainty of biosolids impacts. • Environmental impacts. • Perceived fairness (e.g., site selection). • Transparency of decision-making process. • Economic benefits to farmers, City of Tulsa Public Works and consumers/community at large. • Alternatives to biosolids land application. • Legacy issues (e.g., stink of the Arkansas River).

APPENDIX B-3

TULSA COMMUNITY BIOSOLIDS DIALOGUE PRE-TESTS (DIALOGUE WITH COMMUNITY MEMBERS)

Note: A similar protocol was developed for the Landowner Biosolids Dialogue Pre-Tests

Meeting Agenda

March 9, 2010
5:30 p.m. – 7:45 p.m.

Glenpool Library
730 E. 141st St, Glenpool, OK 74033
(918) 746-5190

Sponsors: Brian Bolte and Cindy Crider – City of Tulsa Public Works Department

Facilitators: Gordon Butte and Sara Eggers – Decision Partners

Community Participants: Six to eight local community members who live nearby agricultural land that has recently had a biosolids land application from the City of Tulsa Public Works Department

Meeting Purpose:

To get community member input into how best to communicate with them about biosolids. We will use this meeting to pre-test a draft presentation and brochure developed to address community interests identified in research conducted by Decision Partners in 2009.

Project Background:

Decision Partners has been working with the City of Tulsa Public Works Department to improve how the City communicates about biosolids. In 2009, the researchers conducted interviews with landowners and community members to learn about their values, beliefs, and questions about biosolids land application. They used the research findings to develop a plan for engaging community members in dialogue about biosolids land application and to create communications materials to support this effort. The next step is to get landowner and community member feedback on our draft plan and the draft materials. These pre-test workshops will help us refine the plan and the materials, which will be used by the City of Tulsa and other water utilities around the United States.

This project has been made possible through a grant provided by the Water Environment and Research Foundation (WERF).

Agenda
March 9, 2010

Time	Topic	Who's Responsible
5:30 p.m. – 5:45 p.m.	**Registration and pick up dinner (provided)**	Sara and Gordon
5:45 p.m.	**Welcome and self-introductions** • Welcome • Round table introductions • Overview of the project • Overview of the agenda • Overview of the community outreach and communications plan	Brian Sara and Gordon
5:55 p.m.	**Community Presentation – Introducing Biosolids**	Sara and Gordon
6:15 p.m.	**Participants' Individual Feedback on the Presentation** (one-on-one discussions between participants and Decision Partners or City of Tulsa staff) • What worked well? • What could be improved? • What questions do you have? Hand out biosolids brochure to participants upon completion of the individual discussion.	Sara, Gordon, Brian and Cindy
7:00 p.m.	**Group Discussion** • Feedback on the Presentation and Brochure • Discuss possible signage and get feedback from participants • Discuss key concepts in the presentation and the brochure and get feedback from participants	Sara and Gordon lead discussion
7:35 p.m.	**Wrap Up and Thank You**	Brian
7:45 p.m.	**Dialogue Session Ends**	

APPENDIX B-4

TULSA CASE STUDY
DIALOGUE PRE-TEST DISCUSSION GUIDE

Individual Discussion Guide

1. **What are your first impressions of this presentation?**

2. **What were some of the things that you took away from this presentation?** *Probe for specifics.*

3. **Did any new thoughts about biosolids come to mind after listening to this presentation?**

4. **What [*other*] questions came to mind?** *Probe for specific questions. Note them, but do not provide any answers at this time. We will address those in the detailed discussion.*

5. **What in the presentation did you find most useful? Why?**

6. **What in the presentation did you find least useful? Why?**

7. **What, if anything, in the presentation raised concerns in your mind?**

8. **Are there any topics that we didn't cover, or that you think we should cover in more detail?**

9. **How useful do think that a dialogue session with a presentation like this one today will be to community members? Why or why not?**

10. **How else might we improve the presentation? Why would this be an improvement?**

11. **Do you have any other comments or questions on the presentation or about biosolids land application more generally?** *Probe for specifics, as time permits. Use back of sheet for notetaking.*

Hand the person a brochure to read in advance of the group discussion. Invite them to get some more refreshments – we'll have a group discussion around XX:XX.

Group Discussion

1) *Opening on presentation.* Thank you for offering your detailed feedback on the **presentation.** Can we ask each of you to share 1 or 2 things that you thought were most important or interesting about the presentation? *{keep brief and high level – we should have gotten much of this from the individual and we'll prompt for specific things below.}*

2) What are your first impressions of the **brochure?**
   a) What were some of the things that you took away from brochure?
   b) Which of these messages is the most important, do you think?
   c) Did it raise any other questions in your mind?
   d) Are there any topics that we didn't cover, or that you think we should cover in more detail?
   e) How useful do think that a pamphlet like this one today will be to community members? Why do you say this?

3) There are a few specific topics raised in the presentation and the brochure that we'd really like to get your thoughts on:

   a) What about the idea that biosolids are a **recycled product?**
      i) What comes to mind when you hear that?
      ii) Does it raise any questions?

   b) What about the **benefits of biosolids to the community?** We talked about biosolids benefitting the local community because it keeps a useful product from being wasted at the landfill, it is more economical than landfill, and it demonstrates our local communities' commitment to environmental responsibility.
      i) What comes to mind when you hear those benefits?
      ii) Does it raise any questions?
      iii) Are there any benefits that are missing?

   c) What about the idea that **biosolids are safe?** We talked about our team's commitment to ensuring that biosolids are produced and applied in a way that is safe for people, and animals and then environment. In the presentation we talked about some of the ways that we ensure safety.
      i) What came to mind when you heard or read that in the presentation?
      ii) Does it raise any questions?
      iii) We talked about how researchers continue to improve our understanding on the safety of biosolids. Does that raise any thoughts or questions?

d) *For landowners:* what about the discussion in the brochure about how we select sites for permitting and applying, depending on a number of criteria *Read from the brochure.*
   i) What came to mind when you heard or read that in the presentation?
   ii) Does it raise any questions?

e) Time permitting (and if not mentioned),
   i) Odor
   ii) Environmental benefits – trapping CO_2 in the environment.
   iii) Others?

4) Let's talk about the **proposed signage** that would be posted a few days before the land application.
   a) How would this be valuable to neighbors in the community?
   b) *Landowners:* Does it raise any specific thoughts to you as a landowner that will have the sign on your land?
   c) What were your first impressions of the signage?
   d) Did it raise any other questions in your mind?
   e) Is there any other information you'd like on this sign?
   f) How else might we improve the sign?
   g) *Show two different versions:* We have some other examples of signs. Any comments? Do you have a preference for one? Why do you say that?

5) *Conclusion Question – High Priority.* We discussed several times in the presentation and pamphlet that our team at the City of Tulsa is **committed to engaging** with landowners, neighbors and other community members about the biosolids that we produce and apply.
   a) What comes to mind when you hear that?
   b) Does it raise any questions?
   c) How do you think that these proposed dialogue meetings, and all of the communication materials will help us better engage with community members? Why would this be helpful?
   d) What could we do to improve?

6) Were there any questions about biosolids that you have that we have not yet addressed? *Time permitting, address questions.*

7) Are there any other comments:

APPENDIX C

PROTOTYPICAL COMMUNICATIONS MATERIALS DEVELOPED FROM TULSA CASE STUDY

C-1 *Biosolids Pamphlet – City of Tulsa*

C-2 *Dialogue Presentation and Speaking Note – City of Tulsa*

C-3 *Technical FAQ - City of Tulsa*

C-4 *Biosolids Land Application Signage – City of Tulsa*

APPENDIX C-1

BIOSOLIDS DIALOGUE PAMPHLET – CITY OF TULSA

For more information about the City of Tulsa Biosolids Program, please visit our website at http://www.cityoftulsa.org/city-services/wastewater/biosolids-marketing-program.aspx .

We would be happy to speak with you further about our Biosolids Program. You may also reach us with any questions and comments. Please contact:

Cindy Crider,
Administrative Operations Supervisor
City of Tulsa
175 E. Second Street, Suite 890, Tulsa, OK 74103

Email: ccrider@cityoftulsa.org
Phone: 918-596-9453

For more information about the City of Tulsa's water and wastewater management and other environmental programs, please visit: http://www.cityoftulsa.org/.

For more information about biosolids land applications, we encourage you to visit the following websites:

The U.S. Environmental Protection Agency (EPA):
http://www.epa.gov/OWM/mtb/biosolids/genqa.htm
The Oklahoma Department of Environmental Quality (DEQ):
http://www.deq.state.ok.us/factsheets/water/Biosolids.pdf
http://www.deq.state.ok.us/factsheets/land/Biosolids.pdf

CITY OF Tulsa
A New Kind of Energy

The Beneficial Use of Biosolids

City of Tulsa
Public Works Department
Biosolids Program

Biosolids are a nutrient-rich, soil supplement that improves land quality by enhancing the fertility of soils. They are typically used on farms and ranchlands.

Incorporating the biosolids into the soil

Did you know:

- Biosolids are a recycled resource made from the treated by-products of wastewater. This wastewater comes from many different sources in a community, such as homes, businesses and schools.
- Applying biosolids on agricultural land is a safe, cost-effective and environmentally-responsible way to recycle wastewater by-products and put them to beneficial use.
- Biosolids are safe for people, animals and the environment when properly produced and applied onto land.
- The City of Tulsa staff follows all required state and federal guidelines and regulations for producing, testing and applying biosolids to ensure that biosolids are used safely.
- Biosolids benefit local farmers and ranchers by improving the soil quality of their land, nourishing their crops and pastures and boosting their fields' productivity. Using biosolids also lowers their costs by reducing the need to buy more expensive chemical fertilizers.
- Every year over 2,500 truckloads of nutrient-rich biosolids are used to enhance local agricultural land instead of being wasted by burying them at the Quarry Landfill.
- Biosolids have been used for over 25 years in our community and for over 50 years across the county.

Q. How can we be sure that biosolids are applied safely?

The City of Tulsa follows all government regulations regarding how biosolids are produced and applied to land. For example, regulations require that biosolids can only be applied to land that is 250 feet away from any residence and 100 feet from any water supply. All City employees and contractors who work with biosolids are properly trained.

In order to ensure that the land is properly and safely managed according to federal regulations, our team also meets with the farmer or rancher before and after any biosolids land application. The City of Tulsa requires that farmer and ranchers commit to following all government regulations on how their land can be used following an application. For example, as an added safety precaution, the land is not to be harvested, grazed by livestock or used by people for at least 30 days after the application. This is to allow for any possible remaining pathogens to die off naturally.

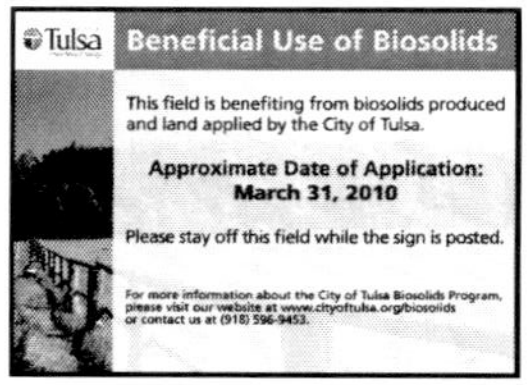

Q. How will I know if biosolids are going to be applied in my area?

We believe it is important to keep neighbors informed about biosolids that are being applied on nearby land. Beginning in 2010, we started a process to let neighbors know about an upcoming biosolids land application. A few weeks before an application, we send information on biosolids to the people who live next to an application site and invite them to attend an informational session hosted by a member of our staff. A few days before the application, a person on our team posts a sign at the site indicating that the field will be applied with biosolids. This sign will remain at the field for at least 30 days after the day the biosolids have been applied.

Q. I've heard that biosolids smell. What do you do to minimize the odor?

Like other fertilizers used to add nutrients to agricultural land, biosolids do have some odor. This odor is typically caused by naturally-occurring sulfur and ammonia, which are nutrients for plants. Typically this odor goes away once the biosolids have had a chance to mix in with the soil, usually a day or two after the application. They may also have an odor if it rains before the biosolids have had a chance to be fully mixed in with the soil, and this too will go away in a day or two. Research on the safety of biosolids has concluded that these odors do not mean that the biosolids present any danger to people or the environment.

Our team is committed to making sure that any odor from the biosolids we produce and apply is as minimal as possible. Certainly, some of the biosolids we produced before 2004 had an unpleasant odor due to the way they were treated in our wastewater treatment plants. In 2004, to address the problems with odor, we began using a treatment process that is just as safe and effective, but produces less odor.

Q. How do biosolids affect the environment?

Research and experience have shown that biosolids are good for the environment. They are biodegradable and release nutrients into the soil to improve soil quality and promote plant growth. Biosolids also contain natural materials that act as a "sponge" to help reduce runoff and soil erosion and increase water conservation. They also trap carbon dioxide (CO_2), which naturally occurs in biosolids, in the soil. This means that the CO_2 will nourish plants rather than be released into the atmosphere. Biosolids can be especially useful to improve poor quality land and turn it back into usable green space, because they add back valuable nutrients that have been depleted and stimulate plant growth. They reduce the need for farmers and ranchers to apply chemical fertilizers. And, applying biosolids to local agricultural land keeps 10,000 tons per year of renewable materials out of landfills.

Research and experience have also shown biosolids to be safe for the environment, when properly produced and applied onto land. The City of Tulsa staff conducts tests to ensure that biosolids meet strict government regulations for safety and quality. That testing ensures that the land does not become over-fertilized and that any chemicals, metals or other potential pollutants in the soil stay well below allowable limits. We also make sure that biosolids are applied at the "agronomic rate" for the field, which means that the amount of nutrients added by the biosolids is just enough to supply the grasses and crops with the nutrients that they can use.

Q. Where are biosolids applied in my area?

Our team applies biosolids to farm and pasture land in Tulsa, Osage, Creek, Muskogee, Okmulgee, Rogers and Wagoner Counties. Biosolids are applied only on land that has been specially permitted by the State of Oklahoma to receive biosolids.

Q. How are biosolids applied?

Before biosolids are applied to a field or pasture, the land site is tested by a City of Tulsa biosolids team member to make sure that the land is suitable for biosolids land application. Our team meets with the farmer or rancher before and after the application to ensure that the land is properly and safely managed according to federal regulations. First, we work with the farmer or rancher to make sure that the ground is prepared to receive the biosolids. During the application, the biosolids are trucked from the City of Tulsa's treatment facility to the application site.

The biosolids are transported in covered trucks by our employees or contractors. Our team then uses special tractors to spread or spray the biosolids onto the ground. Following the application, the farmer or rancher uses a harrow (which is pulled behind a tractor and looks like a rake made out of chains) to incorporate the biosolids evenly into the soil. Our team conducts periodic site inspections and soil tests following biosolids land applications.

Applying the biosolids

City of Tulsa Biosolids Team

The City of Tulsa Public Works staff is responsible for managing the wastewater that is generated by homes, businesses and other buildings in the City of Tulsa. Every day our team makes sure that wastewater – up to 103 million gallons per day – is processed, treated and safe to go back into local streams and rivers. The by-products removed from the wastewater during treatment are further processed into a product called "biosolids" that can be applied on local agricultural land as a beneficial soil supplement.

Our City of Tulsa biosolids team is committed to producing and applying a high-quality biosolids product that benefits farmers and ranchers and meets strict government requirements for safety to people, animals and the environment. We'd like to share with you some information about biosolids. Here are some answers to some of the questions we are frequently asked.

Q. What are biosolids?

Biosolids are a nutrient-rich product used primarily by farmers and ranchers as a beneficial soil supplement. Biosolids are made from wastewater that comes from municipal and septic systems, and comes from many different sources in a community, such as homes, businesses, schools and industries. (The City of Tulsa requires industry facilities to pre-treat their wastewater.) Wastewater is produced whenever people run water – when they flush their toilets or take showers, run dishwashers and washing machines or use garbage disposals. This wastewater contains solids that undergo treatment processes to first remove them from the wastewater and then to make them safe and beneficial for application on agricultural land.

Q. Are biosolids safe?

The simple answer is yes. Research and experience has shown that biosolids present negligible risk to people and environment, when properly produced and applied. Research on biosolids used in land applications continues to improve our understanding of the benefits and risks of biosolids. The City of Tulsa staff follows all applicable state and federal regulations and guidelines for producing, testing and applying biosolids in order to ensure that biosolids are used safely and effectively. We only apply biosolids that meet all of the regulatory standards.

Q. How are biosolids made?

Biosolids are produced at both the City of Tulsa's Northside and Southside wastewater treatment plants. First the wastewater is processed and the solids from the wastewater are removed. These solids, called *sewage sludge*, then undergo a number of processes to destroy potential disease-causing organisms such as bacteria, viruses and parasites, called *pathogens*. Once treated, this product is called *biosolids*. The biosolids may undergo further processing to remove excess water. Depending how much water remains, the biosolids may have the consistency of a thick liquid or soil. The biosolids are stored at the wastewater treatment plant until they can be transported and applied on farm land.

Treatment to remove pathogens

Excess water removed from biosolids

Storing biosolids at the treatment plant

Q. Why does the City of Tulsa produce biosolids?

Everyone who lives, works, shops and plays in and around the City of Tulsa contributes to the production of more than 10,000 tons of sewage sludge per year. As part of its wastewater treatment process, the City of Tulsa is required to treat and manage this sewage sludge. Like many other cities and towns across the U.S., the City of Tulsa staff has found that offering biosolids for use on local agricultural land is the most beneficial and economical way to use this recycled product. This helps local farmers and ranchers, avoids the cost and environmental impacts of landfill disposal and avoids the waste of a useful product.

Q. What does the City of Tulsa do with the biosolids it produces?

Since 1984, our team at the City of Tulsa has provided biosolids to landowners in the surrounding area who have expressed interest in applying biosolids to their land. We select the safest and most economical places to apply biosolids, using criteria such as the distance from residences and water sources, the distance from the wastewater treatment plant, the amount of land available, and the ability of the farmer or rancher to assist in the preparation of the land. Before biosolids can be applied to any land, the City of Tulsa staff must get a permit from the State of Oklahoma. We must show that the land meets strict criteria so we can ensure the safety of people, animals and the environment.

Crop field of land application vs no application

Q. How do biosolids benefit farmers and ranchers?

Biosolids are used by local farmers and ranchers to improve the soil quality of their land and nourish their crops and pastures. The landowners that we work with have told us that using biosolids results in significant and longer-term improvements in crop yield, pasture quality and rate of growth. Using biosolids lowers production costs for farmers and ranchers by reducing the need to buy more expensive chemical fertilizers. Biosolids make excellent fertilizer because they are rich in nutrients such as nitrogen, potassium and calcium. Biosolids also contain organic materials that replenish the soil and boost its ability to absorb and store moisture needed by the crops to grow. These nutrients and other materials are released into the soil slowly and can be used very efficiently by crops. This means that crops benefit from the biosolids for a longer period of time than they would if treated with chemical fertilizers.

Before an application of biosolids

6 months after application

Q. What are the benefits to the community?

Applying biosolids on agricultural land benefits the citizens and taxpayers of the City of Tulsa and its surrounding communities. It is a cost-effective and environmentally-responsible way to recycle wastewater by-products and put them to beneficial use. Every year, the City of Tulsa produces approximately 10,000 tons of biosolids – 2,500 truckloads – that are applied on local agricultural land rather than being sent for disposal to Quarry Landfill! This is good for the environment and reduces costs to taxpayers. And, the success of our biosolids program is a demonstration of our local communities' commitment to environmental stewardship.

APPENDIX C-2

BIOSOLIDS DIALOGUE PRESENTATION
CITY OF TULSA

Slide 1

Thank you all very much for coming. My name is XXXXX and I am the XXXXXX.

Slide 2

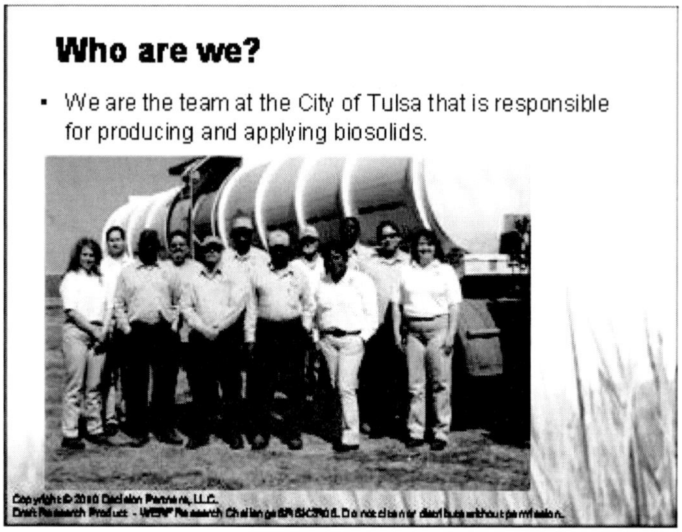

I am a part of the team at the City of Tulsa that is responsible for producing and applying biosolids. We are part of a larger team within the City's Public Works Department that is responsible for making sure that water is safe to drink and that wastewater is safe before going back into local streams and rivers. Our team is made up of 12 staff and contractors, led by XXXXX.

Slide 3

We are scheduled to apply biosolids in a neighboring field in a few weeks. And we believe it is important talk to our neighbors about the biosolids that are being applied on nearby land. Today we'll talk about what biosolids are, why we make them and apply them on agricultural land, and how we make and apply them on agricultural land. At the end, I'd like to hear from you – any questions or comments you may have. Let me know if there is anything that you need help understanding as we go through.

Slide 4

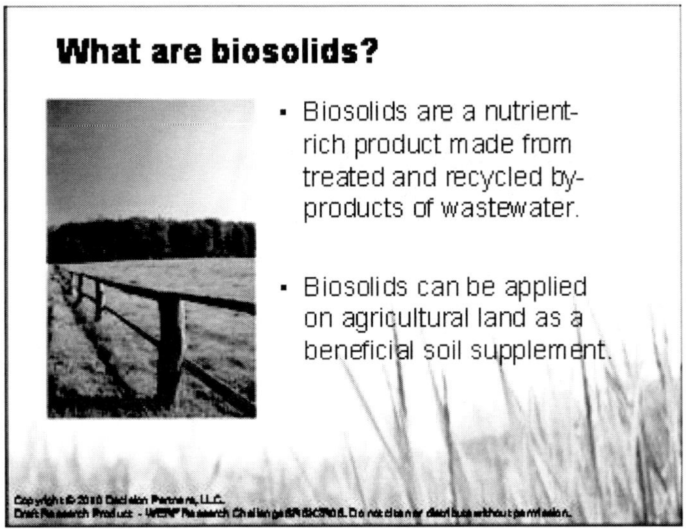

Biosolids are a nutrient-rich product made from treated and recycled by-products of wastewater. They can be applied on agricultural land as a beneficial soil supplement.

We will talk about how biosolids are made and what we mean by soil supplement as we go through this presentation.

Our team has made and applied biosolids on local agricultural land for over 25 years.

Slide 5

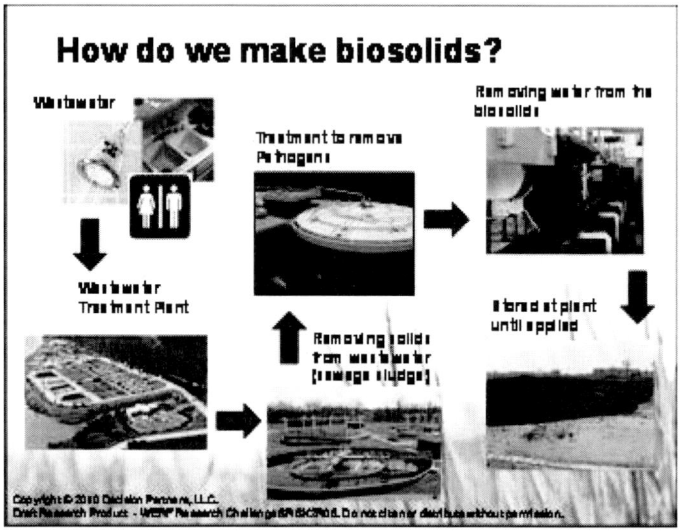

This slide gives you a picture of how biosolids are made here in Tulsa. (*Point to each as you describe.)*

- Biosolids are made from wastewater that comes from homes, businesses, schools and other buildings in our community.
- Wastewater is produced whenever people run water –flushing toilets, taking showers, run dishwashers and washing machines, using garbage disposals.
- This wastewater comes to one of our wastewater treatment plants around Tulsa. Here, the wastewater is processed and the solid by-products are removed. These by-products are called sewage sludge.
- Then, the sewage sludge undergoes a number of treatment processes to destroy potential disease-causing organisms call pathogens. In Tulsa we use a process called anaerobic digestion, which uses natural biological processes like fermentation, to destroy pathogens. These processes also reduce odors. It takes about 20 days to complete treatment. Once treated, the product is called biosolids.
- Some of the biosolids may undergo more processing to remove water. What you see here is called a belt filter press – it essentially squeezes the water out of the biosolids as they travel on a conveyor belt. Depending on how much water we take out, the biosolids may look like a thick liquid or like soil. The biosolids we generally apply here in XXX County has most water removed generally looks like soil, as you see in this picture. (*point to last picture)*
- Biosolids are stored at the treatment plant until they can be transported and applied on farm land.

We produce about 10,000 tons of biosolids every year – that is about 2,500 truckloads.

Slide 6

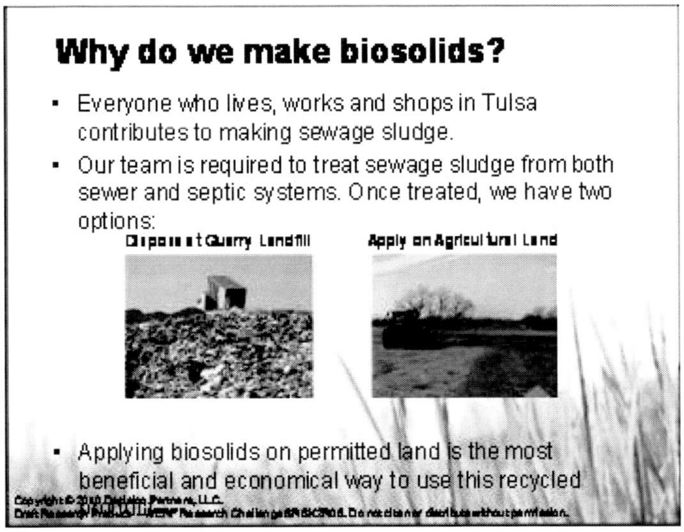

Why do we make biosolids? Well, everyone who lives, works, eats, plays and shops in Tulsa contributes to making sewage sludge. Our team at City of Tulsa is required to treat sewage sludge, both from sewer and septic systems. Once treated, we have two options: We can dispose of it into the Quarry Landfill. Our other option is to offer it to local farmers and ranchers for use on agricultural land that has been specially permitted by the State of Oklahoma.

We have found that applying biosolids on agricultural land is the most beneficial way to use this recycled product. We have offered them to local farmers and ranchers since 1984. Biosolids have been applied to farms, ranches, forests, old mine sites and other types of land across the U.S. for the past 50 years.

Slide 7

Biosolids offer some significant benefits to the local farmers and ranchers that use them.

- They make excellent fertilizer because they are rich in nutrients such as nitrogen, potassium and calcium. These nutrients and other materials are released into the soil slowly and can be used very efficiently by crops. This means that crops benefit from the biosolids for a longer period of time than they would if treated with chemical fertilizers.
- Biosolids also contain organic materials that replenish the soil and boost its ability to absorb and store moisture. This is good for the plants and it also helps prevent soil erosion.
- The landowners that we work with have told us that using biosolids results in significant and longer-term improvements in crop yield, pasture quality and rate of growth. Here is a quote from a local landowner. (*Refer to quote on slide)*
- Using biosolids lowers production costs for farmers and ranchers by reducing the need to buy more expensive commercial fertilizers.

Slide 8

I'd like to show you an example. This is a picture of a field before an application of biosolids.

Slide 9

And here is what that pasture looks like 6 months after the application.

Slide 10

For a more direct comparison, here is a crop field. A few months prior to the planting, the right side of this field had an application of biosolids. The left side did not. You can see the difference in the growth and health of this crop.

Slide 11

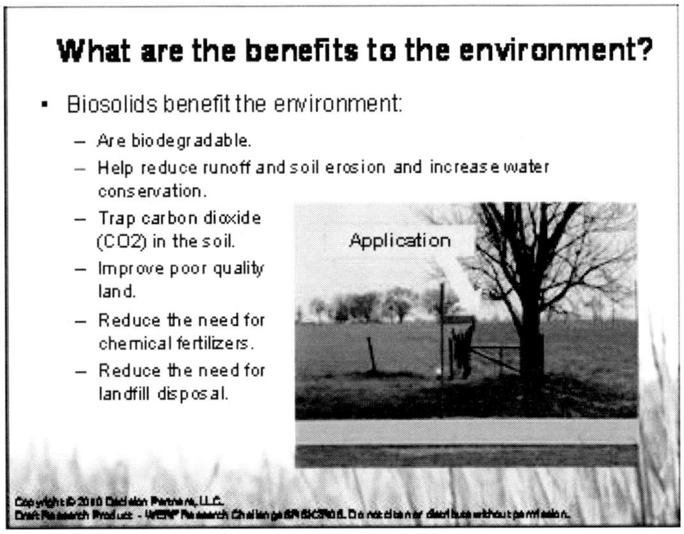

Most of the same benefits that biosolids provide to the farmer and rancher are also benefits they provide to the environment more generally.

- Biosolids are biodegradable. As they break down in the soil they slowly release nutrients that improve soil quality and help plant growth.
- They contain natural materials that act as a sponge to help capture moisture, reduce runoff and soil erosion and increase water conservation.
- They trap carbon dioxide (CO_2), which naturally occurs in biosolids, into the soil so that it is used by plants rather than released into the atmosphere.
- They are especially useful to improve poor quality land and turn it back to usable agricultural land or green space.
- As I mentioned, applying biosolids reduces the need for farmers and ranchers to apply chemical fertilizers, which have their own environmental downsides of runoff and potential leaching into water.
- Applying biosolids also means that we're not disposing this material in the landfill.

Slide 12

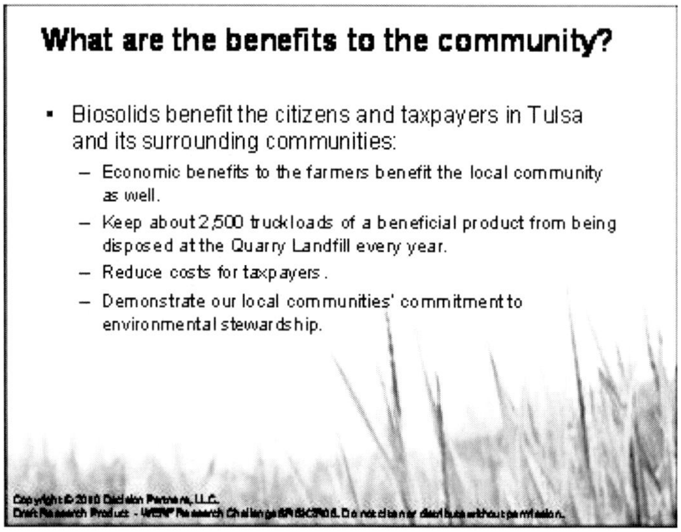

Applying biosolids on agricultural land benefits the citizens and taxpayers of the City of Tulsa and its surrounding communities. It is a cost-effective and environmentally-responsible way to recycle wastewater by-products and put them to beneficial use. It is also good for taxpayers– it costs more to dispose a truckload of biosolids than it does to apply it on local land. Finally, the success of our biosolids program is a demonstration of our local communities' commitment to environmental responsibility.

Slide 13

I'd like to now describe how we apply biosolids – this is how we will do it for your neighboring field in a few weeks. The application will take place approximately March 31, but the exact day depends on things like the weather and ground conditions. Before biosolids are applied to a field or pasture, the land site is tested by a City of Tulsa employee to make sure that the land is suitable for biosolids land application. We also meet with the farmer or rancher before and after the application to ensure that the land is properly and safely managed according to federal regulations.

(Point to each)

1. First, we work with the farmer or rancher to make sure that the ground is prepared to receive the biosolids. Here, you see the farmer aerating the field, which really adds space and air for the biosolids in the soil.
2. During the application, the biosolids are transported in covered trucks, by our employees or contractors, from the treatment facility to the application site.
3. Our employees then use special tractors to spread or spray the biosolids onto the ground. It might be hard to see in this picture, but it looks like small bits of dirt being tossed on the ground.
4. Following the application, the farmer or rancher uses a harrow (which is pulled behind a tractor and looks like a rake made out of chains) to incorporate the biosolids evenly into the soil. *(Note that this isn't the right picture, but we'll get a picture of the right thing in here.)*

Our team conducts periodic site inspections and soil tests following biosolids land application.

Slide 14

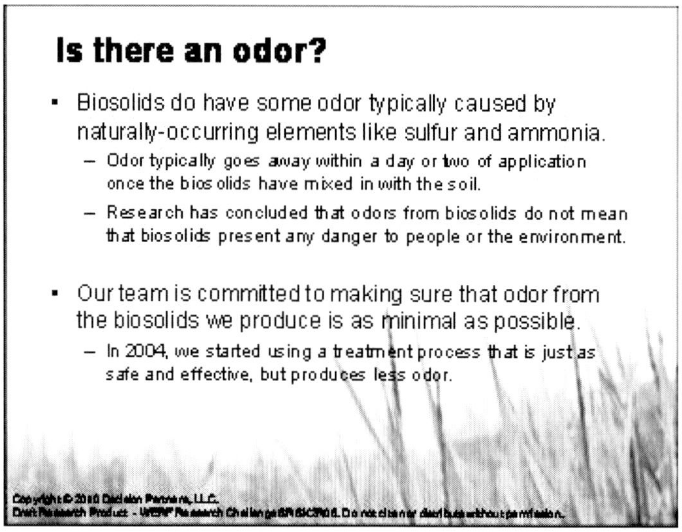

Over the years people have talked with us about odor.

- Like other agricultural soil supplements, biosolids do have some odor. This is typically caused by naturally-occurring sulfur and ammonia. Ammonia contains nitrogen which is good for plant growth.
- Any odor typically goes away within a day or two of application, once the biosolids have mixed in with the soil. Occasionally, if it rains before the biosolids can mix completely into the soil, the odor may return for a day or two.
- Research has concluded that these odors do not mean that biosolids present any danger to people or the environment.

Our team is committed to making sure that any odor from the biosolids we produce and apply is as minimal as possible. Certainly, some of the biosolids we produced before 2004 had an unpleasant odor due to the way they were treated in our processing plants. So, to address that, we began using a treatment process that is just as safe and effective, but produces less odor.

Slide 15

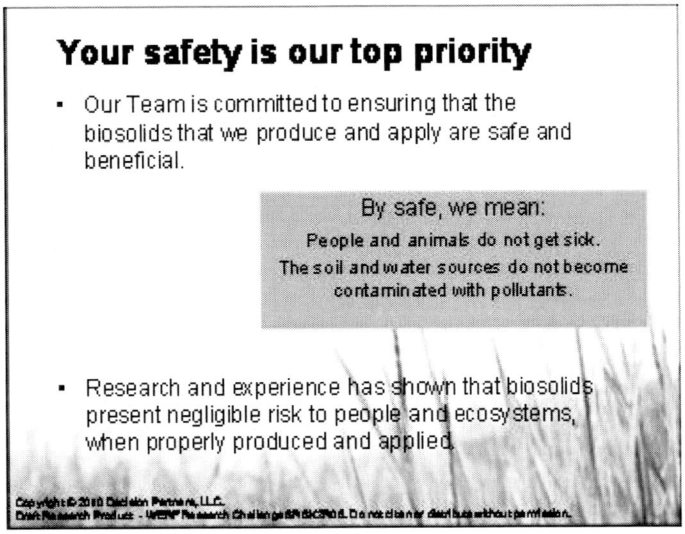

Our team is committed to ensuring that the biosolids that we produce and apply are safe and beneficial. That is, we are committed to making sure that people and animals do not get sick and that the soil and water sources do not get contaminated with things that can pollute it.

Research and experience have shown that biosolids present negligible risk to people and ecosystems, when properly produced and applied. For example, treatment processes are designed to destroy pathogens that naturally occur in wastewater by-products, and regulations and guidelines are designed to make sure that the product is applied only in areas where it should be applied.

Researchers are continuing to study biosolids land application to continuously improve our understanding on the benefits and safety of biosolids.

Slide 16

Here are some of the things that we do to assure safety to people, animals and the environment.

- We follow all government regulations and guidelines for producing, testing and applying biosolids. For example, we make sure that biosolids are not applied to close to people's houses.
- (*not shown*) We stay on top of research and best practices.
- We ensure our team is properly trained.
- We take any biosolids that don't meet standards for safety and quality to the landfill.
- We only apply biosolids on permitted land.
- *(not shown)* We make sure biosolids are applied the "agronomic rate" for the field, which means that the amount of nutrients added by the biosolids is just enough to supply plant with the nutrients that they can use– We do this to make sure that we don't over-fertilize the field.
- We work with the farmer or rancher to ensure that the land is properly and safely managed according to federal standards.
- We conduct site inspections and soil tests.
- We post a sign at the land application site to alert neighbors to a land application site. This sign remains at the field for at least 30 days after the day the biosolids have been applied. (*Remember: this is a pre-test and we don't actually do this yet.)*

Slide 17

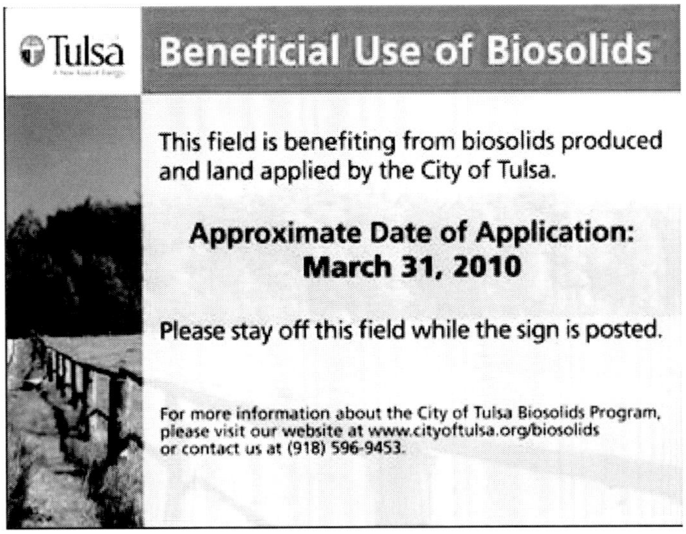

Here is an example of a posted sign. We give the approximate date of the application. We also remind people to stay off the land, and we provide people information on how to reach us with questions.

As an added safety precaution, access to the field is restricted for 30 days after the biosolids application. This is to allow for any possible remaining pathogens in the biosolids to die off naturally.

Slide 18

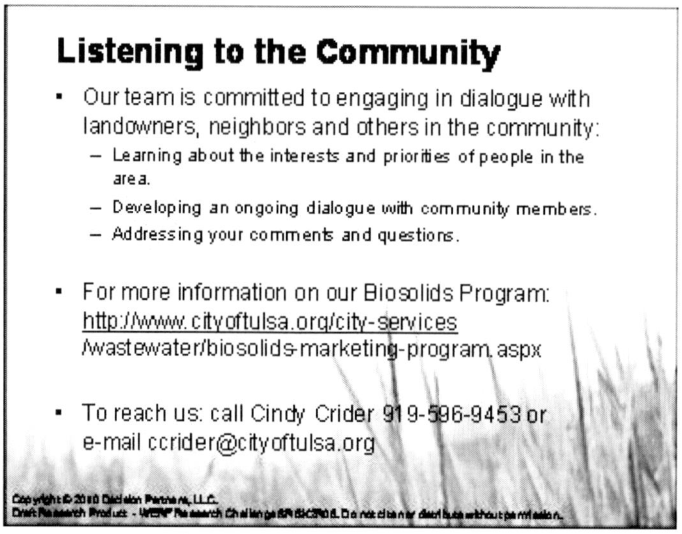

As I've mentioned throughout, we are committed to talking with and listening to you.

- We want to learn about the interests and priorities of the people in the area. Last fall, we conducted some research interviews where we learned in-depth what local residents need to know and want to know about biosolids land application.
- We want to continue this dialogue, and that is why we are developing a plan to have on-going dialogue with you and address your comments and questions. This is why we invited you all to come together today to talk with us about biosolids.
- And we invite you to visit our website (*which we will likely revise in the next few months, thanks to our research)*, or feel free to contact us. Cindy Crider, our Operations Administrator's contact information is here and if you want to ask me anything directly, just let Cindy know.

And with that, I'd like to thank you for listening to me and now and I'd like to listen to you – What questions or comments do you have at this time?

Slide 19

Thank you from our Team!
We'd now like to hear from you – your comments and questions.
Copyright © 2010 Decision Partners, LLC.

APPENDIX C-3

TULSA CASE STUDY TECHNICAL FAQ

Introduction

Using the case study research results as a guide, the Tulsa team developed a brochure and website text, as well as a dialogue presentation and speaking notes, that address many of the key questions neighbors and other community members may have about biosolids. Through the research conducted as part of the case study, more specific questions were raised by the landowners and neighbors who participated in the interviews and in the pre-test dialogue sessions. This Technical FAQ provides Tulsa team members with additional guidance, references, and in some cases draft sample responses, to help them address the more specific questions that may arise during dialogue. Additional references that may be helpful to the Tulsa team are provided at the end of this document.

This Technical FAQ is offered as a resource for WERF subscribers who may be faced with similar questions. **Note**: These questions and sample responses were developed based on the broad key learnings from the research conducted in the Tulsa Case Study. **The sample responses below have not been reviewed by experts. They also have not been pre-tested with community members.** The researchers would recommend that the Tulsa team and any other users, pretest their responses before using them broadly. Guidance for how to pretest simply and efficiently can be found in the *Primer for Biosolids Managers*.

Technical Questions

1. Biosolids land applications:
   - *Additional questions that were raised:*
     - Where is the source of the biosolids used for land applications in our county?
       - *Note: If municipal wastewater or septic system waste from neighboring communities is included in your treatment system, you may want to mention this.*
     - How often do you apply biosolids on a given field? Why?
       - *Note: Be prepared to address whether you apply to different parts of the same field at different times or whether you apply at "half-rates" and why.*

2. Pathogens in biosolids:
   - *Additional questions that were raised:*
     - What types of pathogens are in biosolids?
     - How are pathogens removed from biosolids?
     - How long does it take to remove pathogens from biosolids?
     - What are considered acceptable or safe levels of pathogens in biosolids when they are land applied? Why is this considered safe?
       - *Draft sample response: We, and the regulators who oversee our biosolids operations, work to ensure that the pathogen levels in the biosolids we land apply are as safe or safer to humans, animals and the environment than pathogen levels found naturally in agricultural environments or in*

nature. Our confidence in the safety of our biosolids has been strengthened by a long-term study recently published in the Journal of Environmental Quality, which showed that the process that we use to treat our biosolids eliminates on average 94 - 99% of the pathogens and other organisms they studied from biosolids.*

    - * Pepper, I.L., Brooks, J.P., Sinclair, R.G., Gurian, P.L., Gerba, C.P. (2010). Pathogens and Indicators in United States Class B Biosolids: National and Historic Distributions. Journal of Environmental Quality, 39:2185–2190.

  - ***Note:*** The issues around 'acceptable' and 'safe' are very complex. This sample response should reviewed by biosolids professionals and regulators to ensure that it aligns with expert thinking on biosolids safety. It should then be tested with community members to ensure it is interpreted as intended and that it meets their information priorities and needs. It is recommended that future research, building on the work done in this case study, should explore community stakeholders' interests and priorities regarding biosolids safety.

  - How do you monitor for pathogens in your biosolids?
  - What does the utility do if pathogens are detected in biosolids?

- *Sample resources:* There are a number of research articles and other reports on the pathogen risks. A sample of these references is included in the reference package.
  - The following may be particularly useful to get a broad perspective on pathogens: Pepper, I.L., Brooks, J.P., Gerba, C.P. (2006) Pathogens in Biosolids, *Advances in Agronomy,* 90, pp 1-41.
  - The following 2010 paper is a long-term study demonstrating the effectiveness of pathogen treatment: Pepper, I.L., Brooks, J.P., Sinclair, R.G., Gurian, P.L., Gerba, C.P. (2010). Pathogens and Indicators in United States Class B Biosolids: National and Historic Distributions. *Journal of Environmental Quality*, 39:2185–2190.

3. Other possible contaminants in biosolids:
   - *Additional questions were raised about*: pharmaceuticals, hormones, heavy metals, consumer products, and human diseases (e.g., HIV).
     - ***Note:*** Our research and literature review did not address the safety of biosolids with respect to metals and micro-constituents. Again, communications related to the safety of biosolids are complex and should undergo future research, and most importantly, pre-testing.
     - What other kinds of contaminants and materials does the treatment process address?
   - *Sample references:*
     - 2010 WERF Report: Trace Organic Chemicals in Biosolids-Amended Soils: State-of-the-Science Review (Available to WERF subscribers. Executive Summary found here: http://www.werf.org/AM/CustomSource/Downloads/uGetExecutiveSummary.cfm?File=ES-SRSK5T09.pdf&ContentFileID=15488

- 2010 Water Environment Association of Ontario Report: Assessing the Fate and Significance of Microconstituents and Pathogens in Sewage Biosolids (Available at http://www.weao.org/committees/biosolids/weao-report/Assessing-the-Fate-Report-2010.pdf).

4. Regulations on biosolids land applications:
   - *Note:* Be prepared to discuss the EPA, as well as state and local regulations. Possible website references for the team to review in advance:
     - *water.epa.gov/polwaste/wastewater/treatment/biosolids/genqa.cfm*
     - *water.epa.gov/scitech/wastetech/biosolids/503pe_index.cfm*
     - A Plain English Guide to the EPA Part 503 Biosolids Rule http://water.epa.gov/scitech/wastetech/biosolids/503pe_index.cfm [Note: Although it says "plain English" it does not appear to be written for a non-professional audience.]
   - *Additional questions that were raised:*
     - Why is there a 30-day restriction on access to a field where biosolids (Class B) have been applied?
       - *Draft sample response: The 30-day restriction provides an additional measure of safety, because it allows for the environment, including the sun's ultra violet radiation, heat, wind, and soil microbes to naturally remove any potentially remaining pathogens in the biosolids after treatment.*
     - Buffer zones: Why is there a restriction on how far biosolids must be applied to a residence? Water source?
     - Why is there a restriction that biosolids cannot be applied when it is raining?
       - *Draft sample response: This is an added safety precaution to reduce the chance that newly applied biosolids will run off the field before they have a chance to be fully absorbed into the soil.*
     - How are emerging contaminants accounted for in regulations?
     - How do these regulations compare to those on the use of animal wastes?
       - *Sample reference comparing the risks and benefits of animal wastes:* A 2002 WERF Report Evaluating Risks and Benefits of Soil Amendments Used in Agriculture.
     - Are the state and federal standards reviewed? By whom? When?

5. Wastewater treatment and biosolids treatment processes:
   - *Additional questions that were raised:*
     - What chemicals are used in the treatment process? Are these chemicals also found in biosolids? *Note: Be sure to distinguish between wastewater treatment and biosolids production.*
     - How long does treatment take?
     - What kinds of tests do you do during treatment? Why or why not? Who does the testing?

   Note: Be prepared to discuss more details of the wastewater and biosolids treatment processes. This includes distinguishing between wastewater treatment and biosolids treatment.

6. Biosolids management options:
    - *Additional questions that were raised:*
        - How do other cities and counties like ours handle their biosolids?
            - *Draft sample response: Communities manage their wastewater by-products differently depending on many factors, such as how they process their wastewater in their community, State requirements, their resources and geography. The most common way is how we in Tulsa do it, which is to treat the wastewater by-products so that they can be beneficially used on agricultural land, forest land or reclaimed land such as mines. Some communities, like Milwaukee, WI, treat their biosolids so that they can be used or sold to the public for landscaping and garden use. Some communities, such as Colton CA, are exploring opportunities to convert their by-products to biogas so that it can be used for energy.*
            - For reference, see NEBRA (2007) A National Biosolids Regulation, Quality, End Use and Disposal Survey - Final Report. http://www.nebiosolids.org/uploads/pdf/NtlBiosolidsReport-20July07.pdf.
            - Sample website of other communities. :
                - King County, WA: http://www.kingcounty.gov/environment/wastewater/Biosolids.aspx
                - Layton, Utah: http://www.ndsd.org/
                - Hampton Roads Sanitation District, VA: http://www.hrsd.com/biosolidsprogram.htm - Hampton Roads
                - San Francisco, CA: http://www.carollo.com/Carollo%20Documents/WtrWW_08_BiosolidsManagement.pdf
                - Bay Area Biosolids: http://www.bayareabiosolids.com/
        - What costs are associated with land applications?
        - How much do the alternative methods of disposal cost?

7. Biosolids odor:
    - *Additional questions that were raised:*
        - What causes biosolids to smell?
        - Why do biosolids from one producer/source have a stronger odor than from another?
            - *Draft sample response: A number of factors can affect biosolids odor, including how they are treated, how they are dried, how long they are stored, and how they are applied. Odor will also likely be stronger depending on the weather conditions.*
    - *Note:* Be prepared to address the odors that come from a variety of sources in wastewater treatment, biosolids production (and storage) and biosolids use.
    - Sample references:
        - There are several WERF projects addressing issues of odor. For example, a 2004 Report Health Effects of Biosolids Odors, available through WERF (also found here: http://www.ci.livermore.ca.us/wrd/pdf_files/WERF.pdf).

- EPA has a fact sheet: http://www.epa.gov/owm/mtb/odor_control-biosolids.pdf
- The National Manual of Good Practices for Biosolids, Chapter 6, addresses odor. http://www.wef.org/biosolids/page2.aspx?search=true&searchterm=Odor
- Sally Brown from the University of Washington (http://faculty.washington.edu/slb/) has a non-technical document on odor: http://faculty.washington.edu/slb/docs/basics/Biosolids_and_odor.pdf

8. Land applications decision making:
   - *Additional questions that were raised:*
     - Who decides where to land apply biosolids?
     - What factors are considered when making a decision about where to apply biosolids? *Note: Be prepared to discuss as appropriate:*
       - Preliminary evaluation, including initial site screening.
       - Distance to homes, wells, water bodies, public roads, etc.
       - Environmental factors, such as soil types, texture, drainage and slopes, and physical features of the landscape, soil productivity and crop potential.
       - Distance needed to transport biosolids.
       - Landowner's ability to support with any of the preparatory or follow up work.
       - Nutrient assessment: Ensuring that the application of nutrients is balanced with the needs of the crops to be grown, in order to protect surface and ground water resources from excess nutrients.
     - Do neighbors have a say in whether biosolids are applied on a nearby field?

9. Scientific understanding and research on biosolids:
   - *Additional questions that were raised:*
     - Who is doing research?
       - *Draft sample response: Several organizations including government agencies and those representing the wastewater profession, conduct or support research on biosolids. These include the U.S. Environmental Protection Agency, the National Science Foundation, the Water Environment Research Foundation and the Water Environment Federation. There are several universities whose researchers have contributed extensively to the scientific understanding of biosolids, including the University or Arizona, the University of Utah, the University of Washington and Drexel University.*
     - What are the gaps in our understanding about biosolids? What is less understood?
       - *Draft sample response: Ongoing research studies are working to increase our industry's and our stakeholders' confidence in long-term safety of biosolids. We are always looking for ways to continually improve our programs, for example by developing best management practices for odor. And we want to better understand the range of environmental benefits from biosolids land applications so we can share them with our communities.*

Possible references that may benefit interested stakeholders who request more information (to supplement the Tulsa Biosolids Brochure and Dialogue Presentation):

- North East Biosolids And Residuals Association The Science of Biosolids Recycling (http://www.nebiosolids.org/index.php?page=science).
- EPA factsheet on biosolids: http://water.epa.gov/polwaste/wastewater/treatment/biosolids/index.cfm
- Oklahoma Department of Environmental Quality Fact Sheets:
  - http://www.deq.state.ok.us/factsheets/water/biosolids.pdf
  - http://www.deq.state.ok.us/factsheets/water/Land%20Application.pdf
- Water Environment Federation Fact Sheet: Land Application and Composting of Biosolids (http://www.wef.org/WorkArea/linkit.aspx?LinkIdentifier=id&ItemID=6889)
- Dr. Sally Brown, professor at the University of Washington has put together some brief papers on "The Basics of Biosolids": http://faculty.washington.edu/slb/biosolids_basics.html

Summary of other references that may be helpful for the Tulsa Team (but are not recommended for distribution to stakeholders unless specifically requested)**:**

- EPA (http://water.epa.gov/scitech/wastetech/biosolids/biosolids_index.cfm)
  - See the 2002 National Research Council Report: Biosolids Applied to Land: Advancing Standards and Practices http://www.epa.gov/waterscience/biosolids/nas/complete.pdf
  - See: A Guide to the Biosolids Risk Assessments for the EPA Part 503 Rule: http://water.epa.gov/scitech/wastetech/biosolids/503rule_index.cfm
  - See: A Plain English Guide to the EPA Part 503 Biosolids Rule (http://water.epa.gov/scitech/wastetech/biosolids/503pe_index.cfm)
- Biosolids Associations:
  - National Biosolids Partnership (http://www.wef.org/biosolids/)
    - See: The National Manual of Good Practices for Biosolids.
  - Water Environment Federation (www.wef.org)
  - Water Environment Research Foundation (www.werf.org)
  - North East Biosolids and Residuals Association NEBRA (www.nebiosolids.org)
    - See: *NEBRA (2007) A National Biosolids Regulation, Quality, End Use and Disposal Survey - Final Report.*
  - Mid-Atlantic Biosolids Association (MABA) (http://www.mabiosolids.org/)
  - Northwest Biosolids Management Association (MBMA) http://www.nwbiosolids.org/
- Academic Papers: A number of papers have been included in the reference packet sent directly to the Tulsa team.

APPENDIX C-4

Draft Biosolids Land Application Signage City of Tulsa

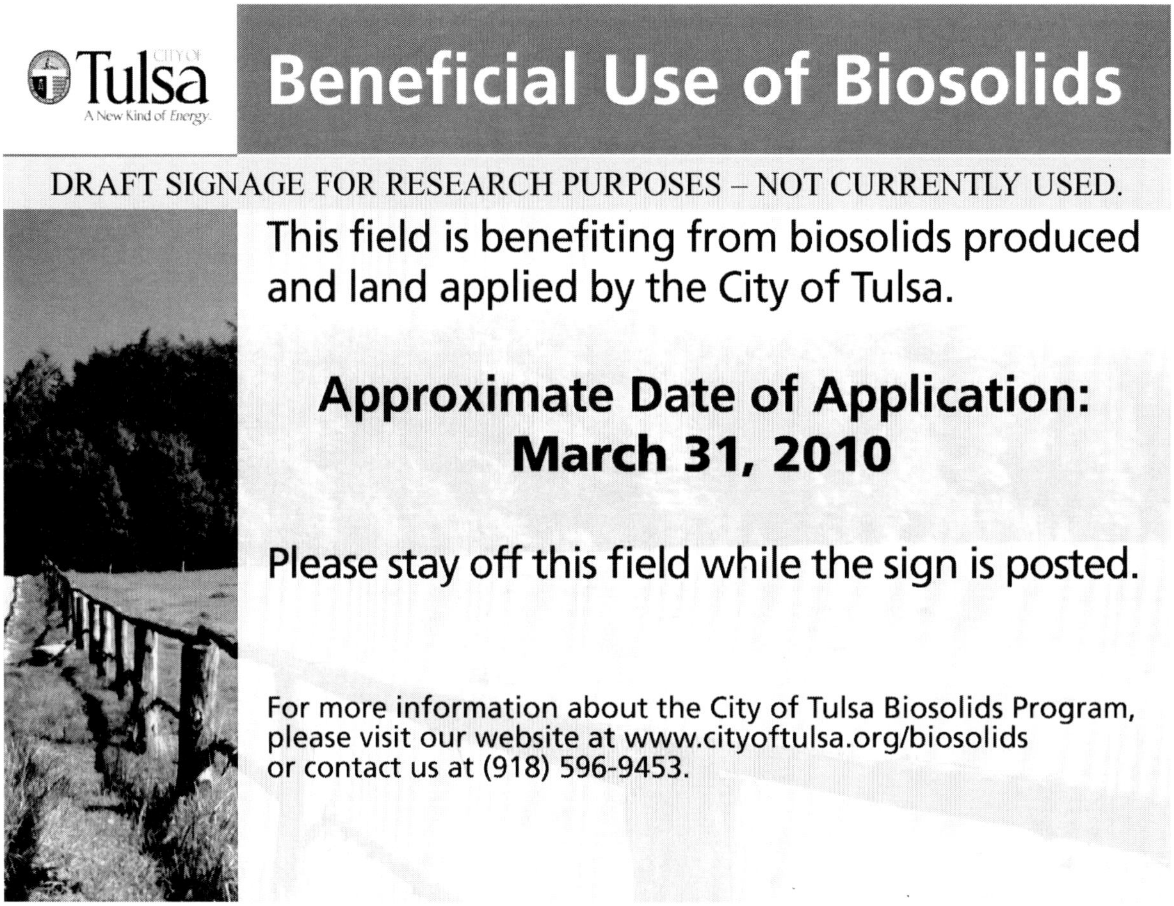

APPENDIX D

VIRGINIA CASE STUDY APPLICATION OF THE STRATEGIC RISK COMMUNICATIONS PROCESS

D.1 Introduction

In 2008, the Water Environment Research Foundation (WERF) and its subscribers recognized the need to develop a systematic, science-based approach to building and conducting effective outreach and dialogue with key stakeholders. As part of a 2008 Biosolids Research Challenge[7], Decision Partners adapted a state-of-the-science Strategic Risk Communications Process[8], and supporting guidance tools and materials that can be used by biosolids professionals and other WERF subscribers to facilitate effective communications with stakeholders, thereby enabling biosolids professionals and their stakeholders to make well-informed decisions and take appropriate actions. This research, described in the main body of this report, should significantly improve WERF subscribers' ability to design and conduct effective risk communications, primarily through outreach and dialogue with their stakeholders – the people in the communities where biosolids are land-applied.

As part of the Research Challenge, Decision Partners adapted and applied the process, supported by its methods and tools, with two case study teams of biosolids professionals. One case study involved a team from the City of Tulsa, Oklahoma, and the other was a team comprising several members of the Virginia Biosolids Council. Based on the insight generated through these two case studies, the researchers developed a Primer (accompanying this report) for biosolids managers to apply the Strategic Risk Communications Process and its methods and tools to conduct more effective dialogue and outreach with community stakeholders in the communities where they operate. This approach will be of particular interest to those responsible for managing biosolids land application and communicating with stakeholders.

This appendix summarizes the case study with the Virginia Biosolids Council (called the VA team). The case study objectives were to apply the Strategic Risk Communications Process in order to:

- Elicit the values, priorities and informational needs of neighbors and regional health officials regarding Virginia communities' biosolids land application.
- Use the research results to support development of risk communications strategies and tools to help VBC members improve their stakeholder engagement and

---

[7] *Through the Water Environment and Research Foundation (WERF) Research Challenge SRSK2R08, Decision Partners collaborated with other researchers[7] to develop methods and tools to assess, evaluate and communicate potential pathogen risks associated with biosolids land application. See WERF Research Challenge SRSK2R08 Final Report for more information.*

[8] *The Strategic Risk Communications Process was developed for Health Canada and the Public Health Agency of Canada (PHAC) by Decision Partners, November 2006. The Process is detailed in the Strategic Risk Communications Framework and Handbook (see www.riskcommunications.gc.ca).*

communications related to biosolids land application, generally, and the safety of biosolids, specifically.

D.1.1 Organization of this Report

This appendix complements this report's accompanying *Primer for Biosolids Professionals: Conducting Effective Community Outreach and Dialogue on Biosolids Land Application* (the Primer). It describes how the Strategic Risk Communications Process was applied in this case and presents the key products, research findings and guidance developed in each step. More detailed information on the Strategic Risk Communications Process, its application in the Case Studies is described in the main body of this report.

The case study is organized against the key steps of the Strategic Risk Communications Process. Following a brief summary of the process, the report describes how each step of the process was applied in the VA Case Study. Research-based recommendations for the VA team to use to improve its community engagements efforts are detailed, including recommendations to:

- Develop and adopt Guiding Principles for Stakeholder Engagement on Biosolids.
- Engage local community members in dialogue.
- Foster effective coordination, communication and cooperation among key partners.
- Ensure effective and consistent messages, including those used on application site signage, the VBC website and the Expert Panel Health Statement.

D.2 Overview of the Strategic Risk Communications Process

Strategic Risk Communications is the purposeful process of skillful interaction supported by appropriate information to enable well-informed decision making and action. The Strategic Risk Communications Process is a step-by-step process to help biosolids professionals design and conduct effective community outreach and dialogue on biosolids land application (Figure D-1). It can be tailored to fit the unique needs of a community or a specific issue or challenge with key community stakeholders. These stakeholders include landowners, people who live near land application sites and other community officials. The process can be scaled to address a specific application, or to provide an approach for an organization that is preparing for land application in many communities. And, it supports continuous learning and continuous improvement.

In the Virginia Case Study, Decision Partners worked with the Virginia biosolids team to conduct the first three steps of the process, and to provide concrete guidance on implementing its last three steps.

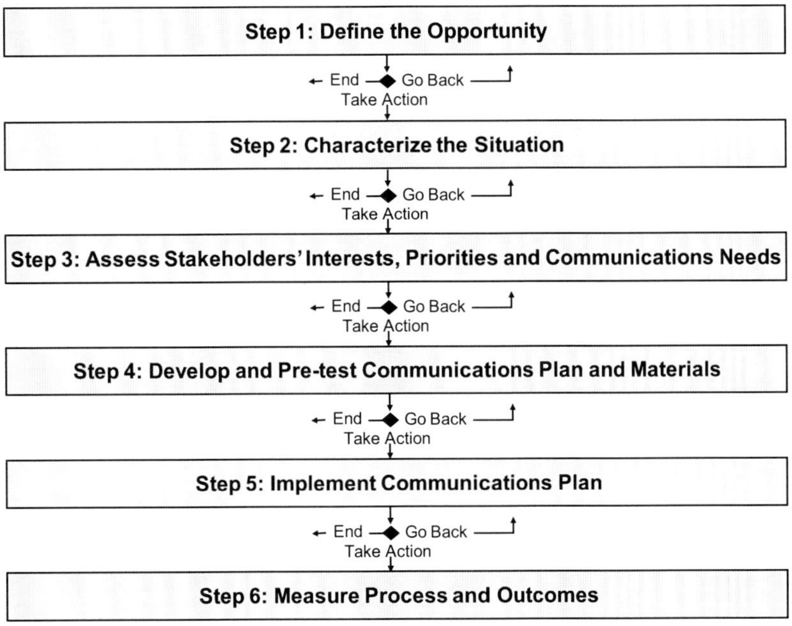

Figure D-1. Strategic Risk Communication Process for Outreach and Dialogue on Biosolids Land Application.

More detailed information on the Strategic Risk Communications Process is provided in Chapters 1.0 and 4.0 of this report and in the accompanying Primer.

D.3 Applying the Strategic Risk Communications Process to the VA Case Study

This section describes how the steps of the Strategic Risk Communications Process were applied for the purposes of conducting community outreach and dialogue on biosolids land application, and describes the key outcomes from those steps.

D.3.1 Step 1: Define the Opportunity

D.3.1.1 Developing a Team

As a first step, Decision Partners and WERF worked with the Virginia Biosolids Council to develop a team made up of biosolids program managers, land application contractors and other practitioners in Virginia:

Virginia Biosolids Council (VBC) members:

- Robert Crockett, Virginia Biosolids Council (VA Team Lead)
- Charles Hooks, Virginia Biosolids Council
- Rhonda Bowen of Hampton Roads Sanitation District (HRSD) (President of VBC)
- Harrison Moody, Recyc Systems
- Hunter Richardson, Synagro
- Chris Peot, District of Columbia Water and Sewer Authority
- Garrett Hughes, (summer intern), District of Columbia Water and Sewer Authority

WERF project team members:

- Alan Hais, WERF
- Sarah Thorne, Sara Eggers and Katherine Sousa of Decision Partners.

The members of the team provided valuable insight and supported the research that was conducted as part of this case study.

The case study was initiated through a workshop with the VA biosolids team on March 5, 2010. The purpose of the workshop was to orient the VA team to the WERF Research Challenge and the Strategic Risk Communications Process and to support the team in adapting and applying the process to the specific challenges and opportunities of the VA team members.

D.3.1.2 Developing an Opportunity Statement

In the workshop, Sarah Thorne and Sara Eggers facilitated the team in developing the following Opportunity Statement, describing the project scope and desired outcomes and defining process deliverables and communications outcomes:

Opportunity Statement

The opportunity of the Virginia biosolids team is to ensure the long-term societal value of our products and services and their sustainability, by systematically engaging and building key stakeholder support for and trust in our people, our products and services, and ensuring effective management of our current and future operations.

As a pilot for the WERF Biosolids Challenge, our work over the next six months will focus on applying the Strategic Risk Communications Process in order to:

- *Discover key stakeholder values, interests and priorities regarding our products and services, particularly as they relate to safety.*
- *Effectively address stakeholders' values, interests and priorities through dialogue, supported by communications tailored to their needs.*
- *Proactively engage key stakeholders in dialogue about the acceptability of our local operations and services.*

D.3.2 Step 2: Characterize the Situation

Step 2 of the process was also conducted as part of the March 5 Workshop. The team participated in a facilitated discussion on key technical, regulatory and other issues related to biosolids land application in Virginia. The VA team members also provided summary documents describing these key issues in their local communities. The team conducted a *Stakeholder Assessment* using worksheets provided by Decision Partners. The purpose of the Assessment was to identify the key stakeholders of biosolids land application in Virginia, and develop hypotheses of their values, priorities, and information needs related to biosolids land application.

The discussions in this step were guided by the **Expert Model of Influences on Integrated Risk Assessment and Risk Communications (RA/RC) System for Biosolids Land Application,** which was developed as part of the WERF Research Challenge. An expert model is a summary of relevant knowledge about a topic, prepared in a way that facilitates risk analysis and risk communications. As a formal integration of knowledge, the expert model summarizes the knowledge needed to make judgments about the topic and issues related to it. The Expert Model developed for this project provides a "systems" view of biosolids land application management, identifying the key stakeholders, processes and potential outcomes in that system. More information on the Expert Model is found in Appendix A.

D.3.2.1 Background on the Communities

As part of the technical analysis, the team described the various communities where biosolids land application occurs in the State of Virginia. Three of these communities were identified as being appropriately representative of communities that biosolids land applicators and managers are likely to encounter:

- **Tidewater:** The Hampton Roads Sanitation District (HSRD) serves several jurisdictions, including, the cities of Virginia Beach and Chesapeake. HSRD runs a "start-to-finish" biosolids program, operating biosolids production and land application. Biosolids are land applied within an approximate 30-mile distance from the utility. The rural areas are quickly becoming more suburban. HSRD has a history of a successful biosolids land application program. It also produces and markets a Class A composted biosolids soil supplement. There is not much local opposition to biosolids land application.
- **Fauquier County:** Fauquier County is primarily rural and agricultural. Nearly all Fauquier's farms are family owned and operated. The county is among Virginia's fastest growing. The county has a long history of biosolids application, with almost 20,000 acres currently permitted. The biosolids that are land applied there are sourced in larger cities in Northern Virginia and District of Columbia and are land applied by companies contracted by the municipal utilities. There is some local opposition to land application, but neighbors seem to be familiar and comfortable with land application.
- **Lunenburg County:** Lunenburg County in South Central Virginia is made up of rural agricultural communities. The biosolids program is relatively new, with less than 5 years experience and approximately 20 permitted sites.

D.3.2.2 Developing a Stakeholder Map

A stakeholder is defined as any individual, group or organization that may affect, be affected by, or perceive itself to be affected by decisions related to biosolids issues, projects, application, problems or policies.

Figure D-2 shows the Stakeholder Map that the Virginia team developed, identifying the stakeholders of a sustainable biosolids business in Virginia. Because the VA opportunity focused on gaining stakeholder trust in the safety of biosolids, the team identified near neighbors and the VA Department of Health (VDH) officials as the priority stakeholders. The team then hypothesized that these stakeholders would view biosolids safety as a top priority and that VDH officials may be a direct source of information for people who inquire about biosolids safety.

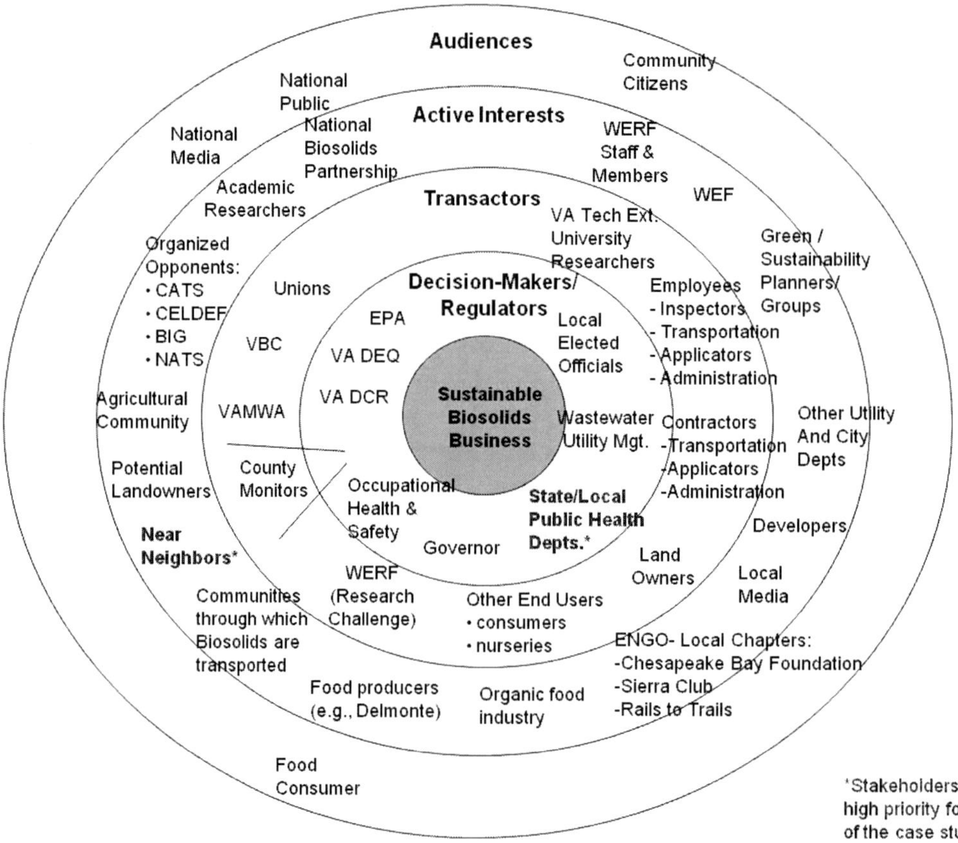

*Stakeholders identified as high priority for the purpose of the case study opportunity.

Figure D-2. VA Biosolids Team: Stakeholder Map of Sustainability Biosolids Business.

D.3.2.3 Developing a Stakeholder Hypotheses

A Stakeholder Hypothesis is used to define who the key stakeholders are and to develop a summary of the team's "best guess" of stakeholders' values, priorities and information needs. The Stakeholder Hypothesis, which informs the development of the research plan in Step 3, is analyzed against the mental models research findings to assess the gaps and alignments. Tables D-1 and D-2 on the following pages summarize the Stakeholder Hypotheses conducted by the VA team.

Table D-1. VA Team Hypothesis on Near Neighbors.

Priority Stakeholder: Near Neighbors	
Assessment Question	**VA Team Hypothesis**
What they know about biosolids and biosolids management	• Biosolids have odor. • Land application results in higher truck traffic. • Biosolids are derived from human waste. • Biosolids contain pathogens.
What don't they know or misunderstand.	• Biosolids build up organic matter and make the land more valuable. • The full range of benefits – reducing greenhouse gas (GHG) emissions, saving landfill space and saving their or somebody's local government money. • Some think biosolids are not safe. • They believe there is a relationship between odor and health effects. • They do not have working knowledge of farming. • They are not aware of the amount of research on biosolids.
What they want to know.	• What is not in biosolids. • What "our" confidence in biosolids is based on. • What is the safety of biosolids. • What to expect in terms of odor and other impacts. • Defined small risks and larger benefits (carbon sequestration, energy use reduction, drought resistance, etc.). • More information on health risks. • The scientific facts, in layman's terms, by a reputable non-biased source. • Times of application so they can plan around them.
Who do they trust?	• DEQ and local government until they think they are getting too cozy with the contractors. • Only trust those who confirm their own opinions. • Citizens in their neighborhood who have done the research on this product and have good stature in the county. • Farmer, Regulatory agency, credible spokespersons. • Someone who is well-educated that they trust (local doctor, local health district professional, county supervisor). • DEQ and the inspectors in the field. • Their neighbor (permitted farmer) and other farmers: they have confidence that he would not do anything to impact his/her land or neighbors. • The information they review on the Internet. • They have less trust in institutions, especially government agencies.
What they value most about biosolids risk management.	• Risk to their health and to their quality of life, their property. • The regulations and buffers in the regulations. • The open space provided by farming, which biosolids helps to preserve. • The need for the farmer to earn a living. • Less concerned with the value that the farmers derive from biosolids. • Research, confirmation of safety, years and experience of use. • Potential for truck traffic. • Odor.
Key influences on their judgments of biosolids.	• Trust factor. • Where the information came from and the credibility of the sources. • What the farmer or land application contractor tells them. • They have been misinformed by activists, the Internet and the media. They have confidence that this information could not be wrong. • They fear unknown constituents that the industry is not telling them about.

Table D-2. VA Team Hypothesis on Health Directors.

Priority Stakeholder: Health Directors	
Assessment Question	**VA Team Hypothesis**
What they know about biosolids and biosolids management.	They know very little about biosolids in Virginia. They know where they come from and that (untreated) they contain pathogens. Most of them are aware of the minimal risks associated with land application of biosolids, and accept that we are not endangering neighbors. However, based on their scientific background, they may be skeptical of the health claims. They usually have to call somebody to get an answer. *How do we know?* There is consistent communication within VDH regarding how to respond to health complaints and perceived health symptoms. There has been some communication between health officials and the environmental regulatory agency regarding biosolids. We had contact with some of them during the DEQ Expert Panel process.
What don't they know or misunderstand.	• There is little if any communication regarding biosolids within VDH. The information they procure, like the general public, is often archaic and confusing. • How safe the product is if done properly, and that it is being done properly. • The science supporting biosolids, for example, the facts about pathogen transmittal and the Risk Assessment process, the facts on bioaerosols in biosolids. • The hierarchy of it, who regulates it, and the oversight provided locally by DEQ. • Some might not know where it is being used, where it comes from. • Some of them assume that if there is a complaint, there is a problem. • They might not understand the benefits from a health perspective (e.g., long-term climate change).
What might they want to know? How do we know?	• The regulations on biosolids land application. • Where the product comes from, how it is tested to ensure its safety, who is responsible for regulatory oversight and ensuring proper measures are in place. • The science, the research, the current practices to protect health and the environment – the scientific facts on pathogens and any research on the issue. • Good, technical, research-oriented information that may help them respond to general alleged health complaints with confidence. • They might want to be briefed regularly on the DCWASA biosolids research program, which looks at some of the persistent chemicals as well as the benefits derived from biosolids.
Who do they trust, and why? How do we know?	• They probably trust other local government officials and state officials in their field. I have experienced it. • Being MDs, they would trust other medical professionals – either one-on-one or in the literature. • Other regulatory agencies. • Other doctors that have conducted peer reviewed research. • I believe they trust DEQ staff and the inspectors.
What communication methods do they prefer? How do we know?	• Meetings with regulators and people they trust. It seems anybody in local, state or federal government disseminates information in a meeting format. • Official conferences held by VDH or other regulatory agencies. Professional organizations. Professional literature. They receive the VDH newsletter, but we have never had any response from them on any issue. • Internal communication among VDH professionals. • They like things in writing – emails and records of complaints and how they were addressed.
What do they value related to biosolids risk management? How do we know?	• Their background in health might give them a comfort level of buffers and regulations because of all the regulations they are familiar with. • They value evaluations of complaints and follow-up, as well as knowing if a potentially affected individual has a pre-existing condition that might mean compromised health. They do not want to risk exposing someone that might be compromised. Therefore, we need to know this ahead of time.
Key influences on their judgments of biosolids.	• They will follow the guidance of their bosses in the bureaucracy first • Given their public health priorities – STDs, low-income health services, etc., biosolids are very low on their list of required reading. They just don't have the time to study each case. This works to our disadvantage.

Priority Stakeholder: Health Directors	
Assessment Question	**VA Team Hypothesis**
	• Biosolids is a political nuisance to them and their likely reaction is to accommodate people by recommending larger buffers for those who claim to have health problems. • They don't have time to weigh the risks and benefits. Some to them might jump to conclusions before they understand the situation. • "How can I get rid of this issue as quickly as possible", which can result in recommending wider buffers. They do not want to have complaints that they are insensitive to health issues. • Quality of research, peer review … it is unlikely a doctor would trust anecdotal information. • Whether a neighbor has a pre-existing condition, regulatory setbacks, history of complaints, etc. We know this because we heard a lot of this during the Expert Panel discussions.

D.3.3 Step 3: Assessing Stakeholders' Priorities, Perceptions, and Communications Needs

From April to July 2010, Decision Partners worked with the VA team to conduct in-depth mental models research with two cohorts: 24 neighbors in three Virginia Communities (Fauquier County, Tidewater and Lunenburg), and six regional VDH officials. This research was conducted as part of Step 3 of the Strategic Risk Communications Process.

D.3.3.1 Background on the Research Method

An expert models/mental models research approach was used to conduct the stakeholder research (Morgan et al., 2002). With its ability to characterize people's interests, perceptions, priorities and communications needs in a structured way, this method can be used to inform the development of communications strategies and materials to address those priorities and needs. It typically involves individual, one-on-one, confidential interviews, either in person or by telephone. The method is well suited to addressing complex risk management and risk communication challenges, which require in-depth understanding of the influences on people's perceptions, judgments, and their resulting decisions and actions. More detail on the method is found in the in the body of the full report (SRSK2R08) and the accompanying *Primer for Biosolids Professionals* (SRSK2R08a).

D.3.3.2 Key Tasks

A mental models interview protocol was developed to solicit near neighbors' and Health Department Professionals' thinking on topics related to biosolids land application, including their values, priorities, perceptions of the benefits and risks associated with biosolids land application, with particular attention to pathogens.

While tailored to each of the two cohorts, the protocol generally followed this outline:

- General thoughts on biosolids, including, why and how biosolids may be land applied, and thoughts after hearing brief descriptions of biosolids.
- Benefits and risks associated with biosolids land application, including potential pathogens in biosolids land application and their impacts to land, water, animals and people. A description was provided.
- Thoughts on the decision-making process for biosolids land application, including how decisions regarding biosolids land application are made, who is involved, and how they are responsible.

- Communications related to biosolids land application, including most important information to share on biosolids land application in communities where they are applied, and thoughts on most effective ways for biosolids managers to communicate about biosolids.

The VA biosolids team next developed a sampling and recruitment plan. The VA team provided geographic coordinates of current or recent land application sites. Decision Partners used Internet-based mapping and reverse address look-up to identify a sample of potential participants. Participants were invited by phone to participate in an interview. Interviewees were offered an honorarium of $30.

D.3.3.3 Mental Models Interviews

Eighteen interviews were conducted with near neighbors in Fauquier County (10) and Tidewater (8) by researchers from Decision Partners between May 18 and June 29, 2010. Six interviews were conducted by a DC Water intern with near neighbors in burg between June 28 and July 27, 2010. See Table D-3 and Figures D-2 and D-3 below for the demographics of the interviewees. Interviews averaged 37 minutes in length, ranging from 24 to 86 minutes. The interviews were audio-recorded with the permission of the participants and transcribed.

Table D-3. Summary of VA Interviewee Demographic Characteristics.

	Near Neighbors	VDH Officials
Number of Interviewees	10 Fauquier, 8 Tidewater, 6 Lunenburg	6
Interview Dates	May 18, 2010-July 27, 2010	June 28, 2010-Aug. 3, 2010
Interview Lengths	Range: 24-86 min. Average: 55 min.	Range: 31-69 min. Average: 45 min.
Number (%) Females	13 (54%)	1 (17%)
Number (%) with Children Under 18	10	NA

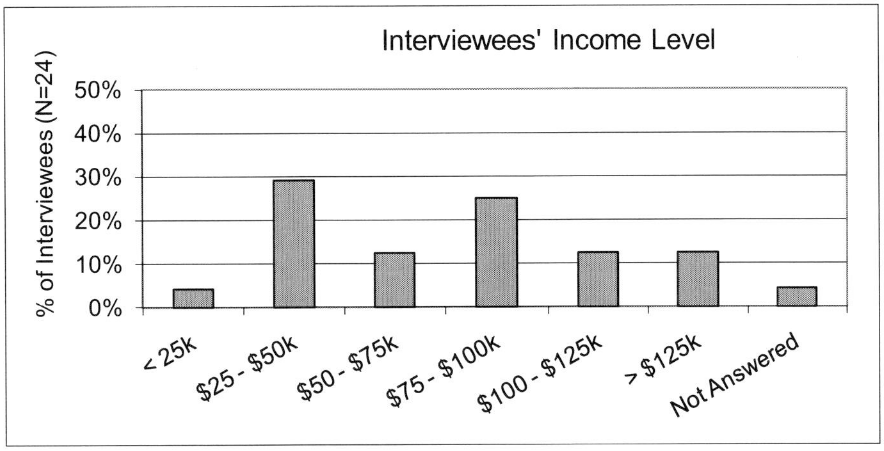

Figure D-3. VA Interviewees' Income Levels

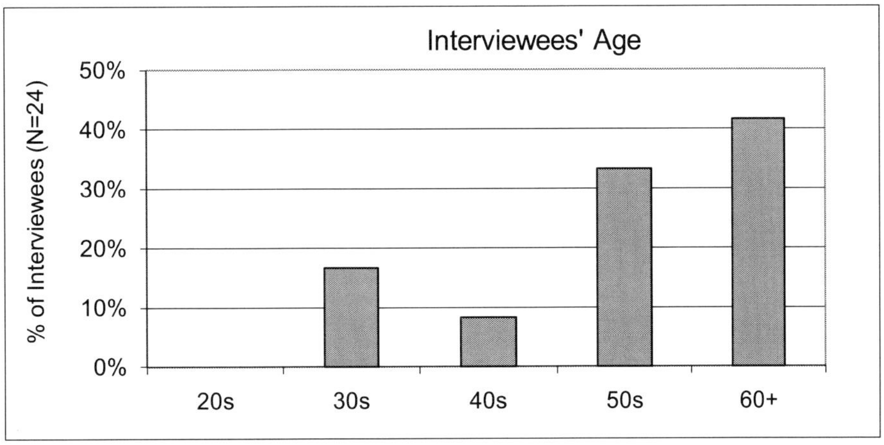

Figure D-4. VA Interviewees' Ages

The interview data was coded and analyzed against the Expert Model and Stakeholder Hypothesis. A detailed research report was presented in workshop format to the team on September 1, 2010. Preliminary considerations on communications opportunities were presented and discussed at that meeting.

D.4 Summary of Key Research Results

D.4.1 Neighbors

The following are the key findings from the mental model research conducted with 24 neighbors of current or recent land application in three Virginia communities (Fauquier County, Tidewater and Lunenburg). These findings are also compared to the *Stakeholder Hypothesis* developed by the VA team in advance of the research.

Most interviewees did not have an in-depth understanding of wastewater treatment and biosolids. Several commented that they were "not a part" of the municipal wastewater system.

- *Comparison to Stakeholder Hypothesis*: This point was not raised by the VA team in the stakeholder hypothesis; however, their feedback suggests that it is an important finding and one that they will address in greater detail in their website and other communications materials.

Interviewees expressed a range of familiarity with biosolids and biosolids land application. Most in Fauquier County were familiar with biosolids and demonstrated basic knowledge of the land application process. A few in Fauquier commented that biosolids is a "controversial topic" in the area. A few commented on having had a negative experience with biosolids (e.g., seeing a spill that had to be cleaned up). Half of interviewees in Lunenburg said they were familiar with biosolids. Interviewees in Tidewater were generally not familiar with biosolids. No one from Tidewater commented on direct experience with biosolids land application (e.g., "haven't paid attention"); however, they speculated that biosolids may be applied on fields near their homes (e.g., by smelling an odor, seeing something sprayed, etc.).

- *Comparison to Stakeholder Hypothesis*: This point was generally consistent with the stakeholder hypothesis except in Tidewater where interviewees demonstrated less familiarity than anticipated.

Interviewees demonstrated a lack of understanding of who is "in charge" of biosolids land application. They spoke in very general terms about decision makers. For example: "local government people", "people who stand to make a profit". Throughout the interview, interviewees offered their judgments of the credibility (both positive and negative) of the people they thought may be key decision makers – the local governments, utilities, biosolids companies, regulators, agricultural extension officials and farmers.

- *Comparison to Stakeholder Hypothesis*: This point was consistent with the VA team's hypothesis regarding who neighbors trust.

When commenting on biosolids, interviewees, particularly those who appeared to be less familiar, often drew on their perceptions of other environmental and health issues. This included their perceptions of how these issues were managed by the people in charge. Examples included: the Gulf oil spill; PCB contamination in the Chesapeake Bay; fly ash and coal ash contamination on public land; animal manure; and commercial fertilizers.

- *Comparison to Stakeholder Hypothesis*: This point was consistent with the VA team's hypothesis regarding what neighbor's don't know or misunderstand about biosolids.

When asked their overall judgments of biosolids (after thinking through the risks and benefits), 50-60% of the interviewees in each community said they were "in favor" of biosolids. (See Figure D-5.)They cited the value to the agricultural community; responsible management of a societal problem; and confidence in biosolids management. About 10% were not in favor, citing the unacceptable uncertainty about health and environmental risks. The remainder (about 30%) said they were undecided or in favor "with conditions". They wanted more assurance about the safety and long-term health and environmental impacts.

- *Comparison to Stakeholder Hypothesis*: This point was consistent with the VA team's hypothesis.

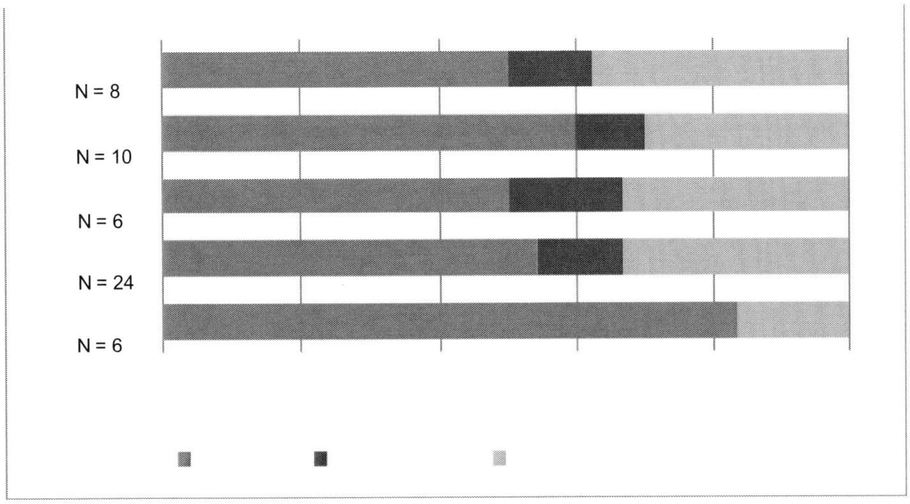

Figure D-5. VA Interviewees' Overall Judgment of Biosolids.

The following appear to be the most significant factors influencing interviewees' judgments of biosolids (in approximate order of priority):

- **Individual's awareness and experience with biosolids** Those interviewees who were more familiar with biosolids land application tended to say that they were "in favor".
- **Confidence in the safety of biosolids** This included the safety of the treatment process (e.g., to remove potential contaminants without introducing other harmful chemicals from treatment), and the safety of the land application. Interviewees who were in favor of biosolids tended to express confidence (e.g., "trusting") that biosolids are safe. Interviewees who expressed less confidence talked about "not seeing enough data" or "being ignorant" of the risks to be able to conclude that biosolids are safe.
- **Concerns about water contamination,** which interviewees believed could result in a number of negative health and environmental impacts.
- **Concerns about the "unknowns" about biosolids** (e.g., about long-term impacts).
- **Trust in the people who produce and apply biosolids** This included interviewees' confidence in the adequacy of the regulations to protect against potential health and environmental impacts.
- **Agricultural and economic benefits** to farmers.
- **Odor** This did not appear to be as significant an issue for neighbors as possible health and environmental impacts. A few, however, particularly those in Fauquier, commented on the impact of the odor on their quality of life and property values.
- Others included: the source of biosolids, use of chemicals, societal need to manage a "human problem".

When read a description about regulations regarding when, where and how biosolids can be applied, many commented that they wanted to know *why* the regulations are in place – that is, whether the regulation indicates some sort of risk. While a few expressed confidence and trust in the regulators, some were less confident, expressing concern about the adequacy of the regulations. Some had questions about buffers, which suggests an opportunity to communicate the purpose of the buffer distances.

- *Comparison to Stakeholder Hypothesis*: This point was not consistent with the team's hypothesis that neighbors value the regulations and buffers in the regulations.

Most Interviewees provided thoughtful comments on the weighing of tradeoffs between the benefits and risks of biosolids. For example, a few who said they were "in favor" of biosolids commented that while there is a smell, it is temporary and an expected "part of farming". A few spontaneously compared the benefits and risks of biosolids to the benefits and risks of using animal manure or chemical fertilizers on land, noting that biosolids are not "any more toxic or dangerous". However, all of those who said they were not in favor of biosolids appeared to have thoughtfully considered the risks and benefits and concluded that the risks do not outweigh the benefits. For example:

> "Opponents of sewage sludge as fertilizer say that there have been studies that say it cannot be thoroughly treated. I think anything can be properly treated. My only concern would be that it be treated in an ecologically sound matter. I don't think treating sewage sludge with hazardous chemicals to kill bacteria may be the answer."

- *Comparison to Stakeholder Hypothesis*: This point is not consistent with the team's hypothesis, which did not acknowledge the neighbors' balancing of benefits and risks when making a judgment of biosolids land application.

Interviewees, particularly those in Fauquier, provided comments on how biosolids managers could increase the acceptability of biosolids land application through their communications and actions. Their suggestions included:

- Addressing and communicating about odor, "otherwise they're going to meet resistance".
- Considering the time of year when making decisions: "Stay away from summer months."
- Demonstrating the value of biosolids through example projects using "damaged areas".
- "Don't mess up."
- Demonstrating its safety before "allowing it to be used on a large scale".

When asked what information on biosolids would be most useful to them, interviewees said they wanted: evidence and/or an assurance of safety; balanced information on the benefits and risks of biosolids application; information on the decision-making process (e.g., where they are going to be used); and information on biosolids treatment and application processes. When asked who should communicate about biosolids, the most common response was government officials, particularly federal agencies such as the EPA or local agricultural extension agents. These officials were believed to have particular expertise and no vested interest. A few mentioned the "biosolids people" (e.g., utilities and companies) because "it's their product". A few mentioned the farmers because they have the knowledge and "can speak to the pros and cons".

- *Comparison to Stakeholder Hypothesis:* This point is generally consistent with the team's hypothesis, although interviewees stressed that they want balanced information on both the benefits and risks, which was not suggested by the team.

When asked how the organizations who produce or use biosolids could best communicate about biosolids, interviewees cited a number of direct, local sources of information, including mailings, newsletters, brochures, email, and "booths at home shows". Local media was also mentioned. Many interviewees, when prompted, indicated that they would like a sign posted in a field before a biosolids application. Few commented on websites, and no interviewees mentioned having seen or read any websites specifically related to biosolids.

- *Comparison to Stakeholder Hypothesis:* This point was generally consistent with team hypothesis, although the team did not specifically suggest neighbors would like signage. The team reacted positively to the recommendation that signage would be a helpful communications tool.

D.4.2 VA Department of Health Officials

The following are the key findings from the mental model research conducted with six officials from the Virginia Department of Health (VDH):

When commenting on the benefits of biosolids, VDH officials generally referred to them in terms of the "best of both worlds" pointing out the agricultural benefits of biosolids and their role in waste management.

- *Comparison to Stakeholder Hypothesis:* This point was not consistent with the team's hypothesis. VDH officials generally appeared to be much more supportive of biosolids than the team anticipated.

Interviewees appeared to be more concerned about emerging contaminants and the "uncertainty of the science" than they were about the pathogens. Some commented that they felt more comfortable with the risks of pathogens: "lots of scrutiny", "proven long-term use", "great deal of experience", "no documented problems", and the "ability to "track" or identify problems. They also focused on the long-term impacts and "non-infectious compounds, such as hormones, medications, heavy metals and "volatile organic compounds".

- *Comparison to Stakeholder Hypothesis:* This point was not consistent with the team's hypotheses. Interviewees' responses demonstrated a deeper level of thoughtful consideration which contradicts their "lack of time to weigh the risks and benefits" and their proclivity to "jump to conclusions before they understand the situation" as suggested by the hypotheses.

A prevalent theme concerned the appropriate management of biosolids: whether adequate regulations are in place and whether those regulations are followed. Four interviewees said they were comfortable with the regulations, citing the stringency of the testing process, the recording of data and DEQ inspection. As one said: When all the rules are followed, it appears to be a safe procedure." Two interviewees were less comfortable with biosolids management, citing lack of requirements for testing, concern about compliance, and addressing emerging contaminants.

- *Comparison to Stakeholder Hypothesis:* This point is generally consistent with the team's hypothesis that VDH officials "might have a comfort level of buffers and regulations because of all the regulations they are familiar with."

Interviewees stressed the importance of community acceptance of biosolids. They believed that "People need to be sold on it." They also believed that land application is a "very big deal to a very motivated, minority of citizens" and "for most people, it's not a particular concern." They believed that the main issues for citizens were the safety of biosolids and "the unknowns"; odor; "not in my backyard"; and transparency.

- *Comparison to Stakeholder Hypothesis:* This point is not consistent with the team's hypothesis, which indicated that VDH officials would not "want to have complaints that they are insensitive to health issues" and would want to "get rid of the issue as quickly as possible". In fact, the research suggests that many VDH officials support biosolids and recognize the importance of community acceptance of biosolids.

Interviewees also said that the following information should be communicated: safety; the treatment process and regulations; application details: where and when biosolids will be applied; information about past and ongoing research; the value of biosolids; and the societal responsibility to manage wastewater by-products.

Many commented that communications on biosolids should come from DEQ or a partnership (DEQ, VDH and utilities). Others suggested that communications should come from researchers. A few believed that communications from those who have a "dog in the fight" (e.g., industry) may be biased or lack credibility.

Interviewees offered thoughtful advice to the biosolids decision makers and their regulators (e.g., DEQ) that focused on demonstrating safety as your top priority and being direct in your assurance of safety – what you know and what you are continuously working to know. For example, one commented about the way DEQ answered questions regarding the safety of biosolids. "[I would say] 'obviously safety for the environment, humans and animals is crucial to us. We are studying every possible lead … to enhance what we've already learned'. I think people want to hear that you're still looking at things."

D.4.3. Bottom Line Research Results

The bottom line conclusions from the research suggest that most people who live in these communities can thoughtfully consider the benefit, risk and cost tradeoffs of biosolids land application in their communities. They can see the value and appreciate the benefits of appropriate and beneficial use of biosolids. They have questions about the ultimate safety and environmental impacts of biosolids and are concerned about what they believe is still unknown. neighbors' confidence in the safety of biosolids rests on their judgments of the credibility and competence of the people who regulate, manage and use biosolids. That being said, they are unclear about who is in charge.

It is clear that communications efforts by the VBC and its member companies have not reached these interviewees. They have not received information through websites, reading materials, or engaged in direct conversation with the people who manage and use biosolids in their communities. Thus, there is an opportunity to develop communications strategies that focus on communicating directly with people living in communities where biosolids land application is, or will be conducted. This communication should be designed to address community members' interests, priorities and questions regarding biosolids land application in their community.

The research also suggests that VDH officials see the value in beneficial use of biosolids in their communities. However, they are not currently engaged by and do not often actively communicate with people about biosolids. The research suggests that there is an opportunity to reach out and engage these officials more frequently and consistently as they have the potential to play a more significant role in communications.

D.5 Research-Based Recommendations for the VA Case Study Team

Using the insight generated from the first four steps of the process, including the research results from the neighbors and the VDH officials, Decision Partners prepared recommendations focused on improving Virginia Biosolids Council's outreach and dialogue with community stakeholders in the communities where VBC members are operating.

These recommendations were developed to focus the VBC and its members' efforts – and future initiatives – on the highest-value opportunities described by the Opportunity Statement. They included guidance on how to effectively engage local community members in dialogue; foster effective coordination, communication and collaboration among key partners; and develop effective and consistent communications materials to support community dialogue. These recommendations can be used to guide the team in completing the remaining Steps 4-6 in the Strategic Risk Communications Process (not conducted as part of this Case Study):

- Step 4 – Develop and pre-test strategies, communications plans and messages: Using the insight generated in the previous steps, the team develops communication strategies, plans and messages that reflect the risk management and risk communications opportunities, and which reflect an understanding of stakeholders' values, interests and information needs.
- Step 5 – Implement risk communications plans: The program manager, with support from the team, implements the plans developed in Step 4. The team monitors and adjusts the plans and messages as required, based on what the team continues to learn through their stakeholder engagement efforts about stakeholders' interests and priorities.
- Step 6 – Evaluate risk communications effectiveness: The team designs and implements evaluation plans for measuring the effectiveness of the Strategic Risk Communications strategies and the effectiveness of the project team process.

Why the focus on engagement with community neighbors? First and foremost, being in such close proximity to land application, they have a significant stake in how that application is managed. Second, research and experience have consistently shown that others often judge the credibility and acceptability of a company or an industry by how they see the near neighbors being treated. This has certainly proven to be the case in the chemical and electricity industries. Therefore, by demonstrating that VBC members are reaching out to the people in the communities where they have operations as a first priority, they will be sending a powerful message about their conduct as a good neighbor and a good community member to the local community and the broader universe of stakeholders.

D.5.1 Develop and Adopt Guiding Principles for Stakeholder Engagement

A key recommendation to the VA biosolids team is to develop Guiding Principles for the Virginia Biosolids Council. While the VBC has a Vision that is clearly stated on its website, it does not have a statement of principles on how its members will engage with the community. Working through the process of developing Guiding Principles that are meaningful to the Council members and making them public is recommended. Adding information to the website that speaks to members' commitment to community neighbors is a specific recommendation.

Guiding Principles are a public statement of what the leaders and employees of an organization (or several organizations in a Council) stand for and how they want their conduct to be judged by their stakeholders. The principles reflect the *character* of the people in the organizations – and what they want to be held accountable for as members of the communities in which they operate.

Why are Guiding Principles important? Because every day these organizations must *earn their social license to operate* in the communities where they are land applying. Like any member of the community, organizations and the people who run them are held to a certain standard by the people in the communities in which they operate. Having Guiding Principles acknowledges their understanding of this responsibility and makes explicit their commitment to be responsible neighbors.

Note: Early in 2010, the VBC had adopted and communicated a Code of Good Practices. This Code was carefully considered when drafting these Guiding Principles for Stakeholder Engagement on Biosolids. The purpose of the draft Guiding Principles is to enhance the Code as it is continually evaluated and refined. The Guiding Principles may also help VBC members identify specific areas of the Code to revise or strengthen.

D.5.1.1 Developing Guiding Principles

There are many ways to develop Guiding Principles, building on an understanding of the people who make up an organization. Often formal research is conducted with leaders and employees. In other cases, workshops are conducted with leaders, principles are drafted, and then workshops are conducted with employees – a critical stakeholder for any organization – to validate the principles and to ensure they are realistic, achievable and credible.

In this case, Decision Partners developed draft Guiding Principles based on our work with the VA team, our assessment of the values and expectations of the people whom we interviewed in the communities for this case study, and our experience with other industry associations. The following draft principles can serve as a starting point for the VBC members to use as they work to develop guiding principles that reflect their values.

Draft Guiding Principles for Stakeholder Engagement on Biosolids
Our Commitment to Our Community Neighbors

As members of the Virginia Biosolids Council, we are proud of our products and services and the value they provide. We strive to be a good neighbor in all the communities that we serve.

We recognize that our decisions, and how we manage our products and services, affect people in the communities where we operate. We know that effective stakeholder engagement and communication is integral to the viability of our operations. We must continue to earn our community neighbors' and other stakeholders' trust in our people, our products and our services.

We do this by:

- *Treating people with respect. We:*
  - *Communicate openly and transparently about our operations and decisions that affect our communities.*
  - *Take people's values interests and priorities into consideration when making decisions that may affect them.*
- *Engaging in dialogue. We:*

- *Seek opportunities to engage people one-on-one or in small groups in ways that build shared understanding.*
- *Provide appropriate and accessible communication channels such as direct inquiry phone lines, and encouraging dialogue through them.*
- *Coordinate with our biosolids partners, including our customers, contractors and landowners to effectively communicate with our neighbors and other key stakeholders.*
- *Actively listen and learn to better understand our stakeholders' values, priorities and informational needs.*
- *Respond to questions and concerns in a timely manner.*
- *Incorporate community feedback when making decisions, where possible.*

- *Being the first and best source of information about our products and services. We:*
  - *Keep our website up-to-date and ensure that the information posted there is useful, relevant and accessible to our Community members and other stakeholders.*
  - *Contribute proactively and constructively to the public discourse on biosolids land application.*
  - *Support community officials in their communications by providing timely, relevant and accessible information.*
- *Recognizing that our actions speak louder than words.*
- *Ensuring that each member of the organizations that make up the VBC demonstrates these values in action.*
- *Continuously learning and improving our operations and communications.*

By living up to these principles, we aim to continue to earn the trust of our neighbors and help assure the sustainability of our operations.

D.5.1.2 Adopting and Implementing the Guiding Principles

If the principles are to be viable and credible, they must be *owned* by the members of the VBC. They also must be doable and achievable. To that end, we recommended the VBC support its members in completing the following steps to draft, approve and implement the principles:

1. Conduct a workshop with Council members to work through this set of draft principles, personalizing and customizing it into something they would feel proud to take back to their respective organizations. For each principle, some key guiding questions should be discussed:
   - What do we do now that demonstrates this principle in action? Can we provide concrete examples?
   - What do our community neighbors and other stakeholders see us doing? Is there a gap between our actions or intended actions and the actions others see?

- What are we doing that may not demonstrate this principle? How could we change that behavior or action? Can we provide concrete examples of what we will do differently?
- How do/will we know that we are living up to this principle? (What would we look for? What would we measure?)

2. Once the principles are refined, each Council member should the share the draft principles (including how they were developed) with the leaders of their respective organizations. This step is important in order to ensure buy-in, commitment and support. The feedback from each leader should be documented – what he or she said and why.
3. With feedback from their respective organizations, the Council members should then meet to finalize the principles and develop a plan for communicating them through their organizations and with their key stakeholders. This includes dialogue, such as at small group meetings, and broadly, as well as VBC website and other print materials. At this time, it may also be valuable to review the VBC Code of Good Practices to ensure alignment.
4. A leader of each member organization should sign off on the Guiding Principles annually on behalf of their companies. This ensures the commitment is evergreen and keeps the focus on the need and desire to live up to the principles. (We suggest it as a discussion topic for the VBC.)
5. Once the Guiding Principles are adopted, they must be effectively implemented by the member organizations leaders and employees. This is essential: adoption of Guiding Principles requires more than just a signature – it requires continual commitment, demonstrated through action. Each principle should be reviewed in the context of the individual member's organization and operations, using the guiding questions listed in Step 1. Individual meetings, workshops and trainings may be required, with support from the VBC.
6. The Guiding Principles – and Code of Good Practices – should be periodically (at least annually) monitored and evaluated to make sure that members' actions and communications align with – and live up to – the stated principles. The guiding questions in Step 1 can support in this evaluation and continual improvement. It may be valuable to include this as a part of annual VBC meetings.

D.5.2 Engage Local Community Members in Dialogue

It is clear from this research that current communications efforts are not effectively reaching community neighbors – at least those in the communities that were part of this research. Thus, there is an opportunity to develop and implement strategies that focus on communicating directly with people about biosolids land application in their neighborhoods. It is important for VBC members to be, and be seen to be, part of the community – as good neighbors.

Communications should not rely solely on a website or brochure. Rather, these tools should support dialogue. A specific recommendation is to design and conduct face-to-face neighborhood dialogue sessions with landowners and their neighbors. This strategy does not mean "town hall". Rather, it is an opportunity to invite landowners' and their neighbors to meet members of the local utility or land application contractor and learn about the specific biosolids land application project in their neighborhood. More importantly, it enables the utility or contractor to learn – and address – neighbors' priorities and questions. As part of the Tulsa Case Study (Appendix B), we developed and pre-tested a plan and supporting dialogue presentation, which may be useful for VBC or its members to develop their own plans and tools.

In addition, the VA team should continue to look for other opportunities to engage with the Community members where they are (e.g., county fairs, local radio or TV programs). There is also an opportunity to engage other key stakeholders, such as the VDH officials and VA Department of Environmental Quality (DEQ) officials in dialogue, through small group discussions, annual meetings and site tours. It is clear that relationships exist with VA DEQ and other state organizations; however, the VA team should look for ways to leverage these relationships with other stakeholders.

D.5.3 Foster Effective Coordination, Communication, and Cooperation Among Key Partners

Effectively communicating with and engaging community stakeholders in dialogue requires effective coordination, communication and cooperation among the key players. We recommended the VA team identify clear roles for each of the key partners, including: the VBC, utilities, contractors, land owners and public officials. The VA team, perhaps working through the VBC should engage with these partners in order to come to agreement on these roles and support their implementation. Mechanisms for clarifying roles and assigning responsibility will need to be established. For example, a contractor's or landowner's role will likely need to be explicitly stated in contracts. Communications processes, materials, training and support will likely also need to be provided.

Table D-4 provides a template for defining these stakeholders and their roles. It can be used by the VBC to identify key communication partners, and specify their roles – the minimum role and the potential higher role – of these partners. Table D-4 also provides a sample of potential roles, based on the case study findings. Table D-4 provides a sample of potential steps that the VBC and its members could take to facilitate partners' communications roles. Note: This table represents only a sample of potential partners and their roles. We suggest VBC work with its members to discuss and refine this table.

Table D-4. Draft Communication Partners' Role in Community Engagement.

Communication Partner	Minimum Role in Communications	Potential Higher Role in Communications
The VBC	• Providing support to members in communications activities. • Maintaining the VBC website. Ensuring it is up-to-date and provides a clear contact where people can get more information. • Handling stakeholder and media inquiries effectively and in a timely manner. • Identifying opportunities to engage community members in dialogue on biosolids.	• Develop and support adoption of Guiding Principles for Community Engagement. • Develop an outreach plan to engage with landowners, VDH and DEQ. • Develop templates and prototypical communications materials that member organizations can customize for their own purposes.

Communication Partner	Minimum Role in Communications	Potential Higher Role in Communications
Utilities	• Maintaining accurate and up-to-date product information. • Providing contractors with information on products. • Notifying VBC of any potential issues that could have communications implications for the VBC and/or the biosolids profession more generally. • Having a website that is up-to-date and clearly identifies how people can get more information. • Having clear communications requirements in their contracts with truck drivers and land application contractors. • Ensuring contractors and landowners have a business card from the utility that they could hand out if needed letting near neighbors know whom to call/what website to look up for more information. • Handling stakeholder and media inquiries effectively and in a timely manner.	• Contacting and/or distributing communications materials to neighbors in advance of an application. • Conducting or participating in neighbor dialogue sessions (e.g., at the landowner's home). • Hosting site tours and/or webinars for key stakeholders (including VDH officials and near neighbors). • Posting tour video on website. • Linking website to VBC website. • Identifying opportunities to engage community members in dialogue on biosolids.
Land Application Contractors	• Displaying their company name, and where appropriate, have a placard with the utility name. • Ensuring professionalism in operations and communications. • Directing inquiries to utility operators in a timely manner. • Notifying VBC of any potential issues or concerns immediately.	• Contacting and /or distributing communications materials to neighbors in advance of an application. • Conducting or participating in neighbor dialogue sessions (e.g., at the landowner's home). • Identifying opportunities to engage community members in dialogue on biosolids.
Landowners who receive biosolids	• Displaying land application signage on site a specified number of days ahead of the application and through the application period as required by contract. • Be willing to speak to neighbors about the use of biosolids on their land. • Providing communications materials to neighbors who inquire. • Ensuring professionalism in operations and communications. • Directing inquiries to specified contact person at the utility immediately. • Notifying specified contact person of any potential issues immediately.	• Contacting and /or distributing communications materials to neighbors in advance of an application. • Hosting neighbor dialogue sessions (e.g., at their home), facilitated by specified contact person at utility or contractor. • Being a credible spokesperson for the utility, contractor or the VBC.
VA Department of Health (VDH) officials*	• Providing communications materials to community members who inquire. • Directing inquiries to specified contact person at the utility immediately.	• Participating in community stakeholder engagement activities hosted by the VBC, utility or contractor. • Identifying opportunities to engage community members in dialogue on biosolids.

* Note: VDH officials were included in the draft Table because they were interviewed as part of our case study research. Additional regulatory and public officials should also be considered, such as VA DEQ officials and local biosolids monitors.

D.5.4 Ensure Effective and Consistent Messages

The case study, supported by research, demonstrates the value in ensuring that the messages and communications efforts of the VBC and its respective member organizations are aligned and coordinated communications. The VBC can support this recommendation by using the research results to develop standard communications materials and templates that can be customized by members to meet their specific needs and those of the people in their communities.

The materials should address the values, interests and priorities of the people in the communities in which VBC members operate. They should be clear, written at a level a layperson could understand (typically, an 8th-grade level), and accessible. Above all, they should be pre-tested with community members to make sure they are understood and interpreted as intended. (Specific guidance on pre-testing is included in the Primer.)

Based on our research we offer the following specific recommendations. VBC and member organizations' materials should:

- Have a consistent definition of biosolids.
- Provide necessary background – context – about wastewater management generally and biosolids management specifically, enabling people to see their "role" in the system – including people who have sewage and septic systems.
- Frame biosolids land application as a beneficial product that, when used appropriately, can yield benefits to people in this community and more broadly.
- Emphasize your commitment to safety as well as environmental responsibility. The VBC website (and others from member companies) clearly states this focus on environmental responsibility; explicit references to the commitment to safety are less clear.
- Address public health officials' and land application neighbors' concern about safety of biosolids – and how you know they are safe.
- Facilitate shared understanding – and build the credibility – of biosolids land application requirements and safe practices, including discussion on the underlying rationale.
- Demonstrate that biosolids production, management and application by VBC members is done appropriately and with respect to the interests and priorities of the people in their communities.
- Clearly communicate about specific projects – what, when, where and for how long, plus how to get more information.

D.5.5 Communicate in Advance of a Land Application through Signage

In working through the case study, the VA team members identified an opportunity to develop a draft template for signage that could be posted before, during and for a specified time after a biosolids land application. The team saw signage as an opportunity to provide more openness and transparency in their operations, and demonstrate their pride in their products and services. They believed – supported by the research results – that signage would have a great deal of merit in their communities, and be appreciated by neighbors. In addition, a proactive step in developing and using signage could be an important voluntary initiative that could either a) offset the need for regulated signage or b) be accepted by lawmakers as appropriate signage.

The team identified a number of considerations on the development of the signage. They stressed the need to anticipate the likely regulatory requirements that may be mandated in the future. The team also stressed the importance of efficiency and cost-effectiveness – signage that can be mass-produced and easily modified. Another consideration was enabling the buy-in from and coordination with the landowner or operator of the field site.

Based on the research results, and the Tulsa Case Study[9], the following text was drafted as a starting point for developing signage for local application by VBC members. It will be important to pre-test the signage with landowners and their neighbors before implementing any signage strategy. It may also be useful to engaged lawmakers and others in discussion at the conceptual stage, to get their input.

Draft Signage Text

[Include Organization Logo (Utility, Contractor or Both)]

A Beneficial Use of Biosolids

This state-approved site is benefitting from biosolids produced and applied as part of a comprehensive nutrient management plan.

As an added measure of safety, please keep off this land while this sign is posted.

For more information, please contact our office at [Phone #].
Or visit: www.virginiabiosolids.com

[Include State of Virginia Permit Number]

D.5.6 Reframe the VBC Website

The Virginia Biosolids Council website is an important tool for communicating with stakeholders about biosolids. It is also important for systematically building the public record. The current website demonstrates the VBC's commitment to safety in the Vision statement, which is featured on the home page. The site provides comprehensive information on: biosolids and biosolids land application; their benefits and potential risks; research and current knowledge on the safety and environmental impacts of biosolids; and the regulations that guide biosolids land application management, among others. The website is professionally-designed, and relatively easy to navigate.

However, the research suggests that the VBC website is not as effective a tool as it could be for directly engaging local community members or for providing the information they want and need in order to make a well-informed judgment about the industry or the VBC.

We recommended reframing the purpose, content and style of the website to connect at a personal level with local community members and stakeholders. Using the research results as a base, we recommended reframing the website to:

---

[9] *Draft prototypical signage is also presented in the Tulsa Case Study Executive Summary (See Appendices B and C).*

- Be personal, write in voice of the people comprising the VBC, and focus on who the members are and how they want to be seen in their communities.
- Clearly reflect the Vision (as well as the Code of Good Practices and any Guiding Principles for Stakeholder Engagement) throughout the website.
- Focus on the people, operations, products and services. For example, include more pictures of local applications – and the people involved, including utility operators, contractors, and farmers.
- Emphasize the topics of most importance to interviewees, framed as "what you have told us you want us to talk about", which as suggested by the research, include long-term safety of people and the environment, and the land application process.
- Use language of assurance throughout – be positive about what we are doing.
- Avoid appearing to be defensive or argumentative. Be objective, neutral and respectful of the other's underlying values, interests and concerns.
- Be designed and written at a level that is appropriate and accessible for community members, and avoids technical jargon.
- Enable dialogue – it should provide an opportunity for visitors to provide feedback, ask questions, make comments or seek more information.
- If necessary, provide a password protected site on the website for information that is for members only (e.g., technical discussions).

From these general recommendations on the website, some specific recommendations were offered to the team, drawing directly from the mental models research findings with VA neighbors and VDH officials. **Note:** Any suggested changes should be pre-tested to ensure they are interpreted as intended. The following is a sample of those recommendations:

1. Wastewater system diagram: The research suggests that people do not have a clear understanding of the wastewater treatment process. In particular, rural neighbors did not view themselves (or their septic systems) as part of the wastewater system, or as a contributor to biosolids production. Include sewage and septic systems as a source of wastewater or biosolids, and more specific information on the treatment processes for biosolids. If possible, make the diagram as specific as possible to actual communities, for example, through photos of actual treatment facility processes (and if possible, showing people working at those processes.)
2. Benefits to the community: This section could also include many benefits that align with what interviewees indicated were their top priorities, such as soil quality, and the societal benefits including "maintaining a rural way of life." Use photos as tangible evidence of the benefits.
3. Societal Management: Frame biosolids as a beneficial product, a part of our "societal responsibility to manage wastewater by-products". Be careful, however, not to present biosolids as a "solution to a problem", but rather as a "win-win" situation.
4. Biosolids Management Options: Include a section titled "Management Options". Consider including a discussion comparing the environmental and economic benefits of biosolids land application versus incineration and landfill.

5. Addressing emerging contaminants and long-term impacts: Include information on the site that would address stakeholders' concerns about "emerging contaminants", (i.e., hormones, medications). Continue to stress that ensuring the safety of biosolids land application – including emerging issues of concern – is a top priority for your organization. Discuss how you (and others) are continuing to work to address the unknowns and manage risks. Provide links to other credible third party sources doing work in this area.
6. Odor: Include a discussion on the priority that your organization places on minimizing the potential impacts of odors. Include a discussion on "what to expect" in terms of odor, including how odor dissipates over a period of days of time, and why biosolids from one facility might have a stronger odor than one from another.

D.5.7 Reframe the Panel Report

The VA team identified the research as a specific opportunity to evaluate a concluding statement of an expert panel, established by the Virginia General Assembly, about the safety of biosolids. This report is described in an article on the VBC's website ("Biosolids Safe Says Expert Panel"; http://www.virginiabiosolids.com/news/expert_panel.htm). It has also been highlighted in recent presentations delivered by VBC and its members.

As part of the research, interviewees were read the following statement and asked to comment:

"A panel of experts was established by the Virginia General Assembly to study the impact of land application of biosolids on human health and the environment. In its final report, the panel said it uncovered no evidence or literature verifying a causal link between biosolids and illness, recognizing there are currently gaps in the science and knowledge surrounding the issues."

A few interviewees, particularly those in the Fauquier County, expressed confidence in the statement. These interviewees appeared to conclude that biosolids are safe. They tended to place more trust in the regulators and biosolids managers more generally. More interviewees, however, responded with less confidence. They focused specifically on the gaps in knowledge. They also questioned the credibility or neutrality of the panel, saying: "It's a political statement". One interviewee said the statement reduced her confidence in biosolids: "If you don't know about the effects, then that bothers me."

Given these results, it is recommended that the VA team amend the manner in which the VBC summarized the findings. The report serves as credible and concrete evidence in support of a conclusion that appropriate use of biosolids is safe. A revised statement should:

- Focus first on what the study did find rather than its limitations.
- Use language of assurance that reminds people what is not known.
- Be direct, and to the point, and be written at a level that an eighth-grade student could understand.

As a higher-value opportunity, we recommended archiving the original article with new, reframed – and pre-tested – information summarizing the findings of the expert panel. This article could include information on the safety of VBC members' operations, using the expert panel report as evidence (not the main story). This will likely help community neighbors – who hold safety as a top priority and have real questions and concerns – increase their confidence in the safety of biosolids and their trust of the VBC and its members as credible organizations. The article should be written from the voice of the VBC and its members. It should first and foremost re-affirm that safety is the top priority of the VBC members – respectfully acknowledging that it is a top priority of people in the communities where members operate and a value VBC shares with its community neighbors. It should also describe how VBC members (and others) are continuing to work to address the knowledge gaps and appropriately manage risks, and how the VBC is committed to sharing new knowledge with Community members as it becomes available.

D.6 Conclusion and Next Steps

At its Annual Meeting, the VBC members reviewed the research findings and resulting recommendations. The VBC decided to adopt the recommendations, and committed to implementing several action items in its 2011 workplan. As an example, the VBC has already begun redesigning its website to reflect the research learnings and recommendations. The redesigned website should be implemented in 2011.The VBC members also expressed an interest in leveraging opportunities to more effectively communicate with community stakeholders, as recommended in the findings.

The VBC members expressed value in the Strategic Risk Communications Process and the key research learning and recommendations that came out of that process. As part of its Annual Meeting, Sara Eggers, the Research Challenge Principal Investigator, and Alan Hais, the WERF Program Director, were invited to present a summary of the case study to the VBC membership. Following the case study presentation, discussions focused on the importance of focused stakeholder engagement efforts at the community level and the value of the in-depth research conducted with people in the local communities.

Two VBC members on the VA team offered their thoughts on the case study's value:

> "The paradigm in our industry has often been that what we do is extremely important: we are environmental professionals. [District of Columbia Water] is a big operation that affects a lot of people, and we would be a lot more comfortable if the receiving community had a better understanding of who we are and what we do. We found in this project that there are some gaps.
>
> The key take-away for me, through this project, is that proactive, direct communication is the goal. This research project produced tools, techniques and methods to better communicate the risks and benefits directly to the public. We already do some communication and we're going to use that as a framework and a building block to take the tools we've learned from this project and integrate it in to our current communications program."
>
> *Chris Peot, District of Columbia Water*

> "Based on the information we gained in this project, we can further refine our messages to address the issues and concerns as well as the positive comments that we received from the research. Now, we want to take this information and adapt it to our own situation and develop messages that are relevant to our own communities. The information has also been very informative for us to be able to move forward with the communications initiatives as we address new regulations in Virginia."
>
> *Rhonda Bowen, Hampton Roads Sanitation District*

APPENDIX E

MENTAL MODELS RESEARCH PROTOCOL

Solicitation

Hi, my name is <name>. I'm a researcher with Decision Partners. Our company has been asked to conduct research on behalf of the City of Tulsa and the Water Environment and Research Foundation or WERF. WERF is an independent scientific research organization dedicated to understanding wastewater and stormwater issues. We are speaking to people in your area to learn about their thoughts on wastewater treatment by-products. This research will be used by the City of Tulsa and WERF to improve communications with people about these by-products.

I'm calling to ask if you would be willing to participate in a telephone interview. The interview should take about 30-40 minutes and doesn't require you to have any particular knowledge about this topic. We can talk now, if you like, or we can schedule a time that is better for you.

Would you be interested in participating?

- If no: May I ask why you are not interested in participating? Thank you for your time.

- If yes: Thank you. Would now be a good time?
    - If yes: Proceed with **Introduction**
    - If no: Proceed with scheduling at the interviewees' convenience. Begin scheduled call with interview script that starts under the heading: **If calling back at a scheduled time:**

Interviewers note:

- *If people say they don't think they know enough about the subject to participate, say:* The interview is very general and doesn't require you to know anything specific. What we are interested in is people's first impressions, even if they don't know a lot about the topic. If there are any questions that you can't answer, we can just skip over them.

- *If people ask how they were selected to be interviewed, say:* We randomly selected your name from a list of names of people who live in communities around Tulsa. This list was generated with the help of people who work at the City of Tulsa Public Works Department

- *If people ask who is involved in this research, say:* We are talking to people who live in communities around Tulsa.

If calling back at a scheduled time:

Hi, this is << name >> from Decision Partners, calling to keep our scheduled interview. As I mentioned when I set up the interview, we're talking to people in the area to learn their thoughts about wastewater treatment issues. As I mentioned, our conversation should take about 30-40 minutes.

Introduction

Thank you for agreeing to participate. I have some questions to help guide our discussion, but please feel free to raise any topic that comes to mind as we go along. There are no right or wrong answers in this kind of an interview, and all of the comments you provide will add value to the research. I want to assure you that what you say will be kept confidential to our research team. You will not be identified in our report.

Before we start, I'd like your permission to tape record our conversation in order to better prepare my notes afterwards. But as I said, everything you say will be kept confidential. May we proceed on that basis? Thank you.

Interviewer Notes:

- *Be sure to prompt people with follow up questions to find out the "why" of their thinking:*
  - Please tell me more about that?
  - Can you share your thinking on that?

- *Do not provide any information to interviewees other than what is provided in notes or otherwise instructed.*
  - *Several questions ask interviewees to think of questions that come to mind. Note questions, but do not offer any additional information. If they ask you for an answer, say:* I can't provide any additional information at this point but I will take down your question. At the end of the interview, I will give you information about where you can go to get your question(s) addressed.
  - *If needed, remind interviewee that we are interested in what kinds of questions they have about this topic so as to improve future communications with people like them.*

Part 1: General Thinking on Biosolids

I would like to start with a few general questions.

1. What first comes to mind when you think about municipal wastewater?
   Prompt if not mentioned:
   - Do you think it is treated? If so, how?
   - Why do you think wastewater is treated?
   - What do you think happens to the water after it is treated?
   - What do you think happens to the material that is cleaned out of the wastewater?

2. Are you familiar with the term "biosolids"?
   If yes *(or* I think so*):*
   - What can you tell me about biosolids?
   - Where have you heard about biosolids?

If no *(or,* I'm not sure*):*

- If you had to guess, what comes to mind when you hear the term biosolids?

3. Biosolids are also sometimes called *treated sewage sludge*. Are you familiar with this term?
   - What comes to mind when you think about treated sewage sludge?

Interviewer read: Just to make sure we are on the same page, I'm going to read a brief description about biosolids. I'll read a little at a time and then get your thoughts or questions.

Description A: When wastewater is processed at a wastewater treatment plant, human waste and other materials are removed. The materials that are removed are called sewage sludge. Municipal wastewater facilities treat sewage sludge according to government regulations to reduce or remove living organisms, such as germs and bacteria. Treated sewage sludge is referred to as biosolids.

4. Having heard that description, does anything else come to mind?
   - Do any additional questions come to mind?

   Interviewer: Probe for the following:
   - Why might wastewater facilities treat sewage sludge to produce biosolids?
   - What do you think wastewater treatment plants do to treat sewage sludge and produce biosolids?
   - Which term – treated sewage sludge or biosolids – do you think is better? Why?

Description B: Okay, I'll continue with the description. Biosolids are sometimes applied to certain types of agricultural land, and this is called '*biosolids land application.*' Depending on how biosolids are treated, there may also be regulations on when, where and how they can be applied, such as making sure they are a minimum distance away from people's houses or water sources.

5. Having heard that description, does anything else come to mind?
   - Do any additional questions come to mind?

   Interviewer: Q5 may elicit responses that relate to Q6-Q7. Tailor follow-up prompts and questions accordingly.

6. Why do you think biosolids are used in land application?
   - What kinds of agricultural land might biosolids be used on?
   - What do you think happens to biosolids if they are not applied to land?

7. How do you think biosolids are applied to land?
   - Who do you think applies the biosolids to the land?
   - How do you think the biosolids are transported to the places where they are applied?

Description C: Now, I'm going to describe one way that biosolids may be used. Biosolids can be applied to increase nutrients to farmland, such as hayfields and pastures. Currently, the City of

Tulsa's municipal wastewater facility produces biosolids and they are applied in this way to a portion of land in Tulsa, Osage and Creek Counties. For this application, biosolids are typically made into a semi-solid, like mud. The biosolids are transported to the land site by truck. The biosolids are sprayed onto the ground and then mixed in with the soil. Government regulations require that the type of biosolids currently made by the City of Tulsa be applied only to land that is 250 feet away from any residence or 100 feet from any water supply and to land that will not be harvested, grazed or used by people for at least 30 days.

8. Having heard that description, does anything else come to mind?
   Do any additional questions come to mind?
   Interviewer: Q8 may elicit responses that relate to Q9-Q11. Tailor follow-up prompts and questions accordingly.

9. Have you seen or heard anything like this happening in fields near your house?
   - *If yes:* Please describe what you've seen or heard. *{Probe in detail.}*

10. What do you think happens to the land over time after the biosolids have been applied?
    - How do you think environmental conditions, like weather, might affect this?

11. Are you familiar with other types of ways that nutrients are added to farmland in your area? Please describe. *If an example is needed:* These are sometimes called fertilizers.
    - *If not mentioned:* Are there any fertilizers from animals that are used?
    - *If not mentioned*: Are there any fertilizers that are made from chemicals?

Part 2: Impacts Associated with Biosolids Land Application

I'm now going to ask you to think about the possible impacts, both good and bad, that may come from biosolids land application in your County.

***Interviewer**, Note for Qs 12-15:*

- *If* "health" *or* "health impacts" *is mentioned, prompt:* What are you specifically thinking of when you think of health?
- *If* "environment" *or* "environmental impacts" *is mentioned, prompt:* What are you specifically thinking of when you think of the environment?
- *If* "animals" *is mentioned, prompt:* What types of animals are you specifically thinking of?

12. What are the possible benefits of biosolids land application in your county?
    - Are there any other possible benefits that come to mind?

    For each:
    - Who might benefit the most by this?
    - How important is this possible benefit? Would you say it is very important, somewhat important or not very important? Why?

13. What are the possible negative impacts or risks of biosolids land application in your county?
    - Are there other possible negative impacts that come to mind?

    For each:

- Who might be most affected by this?
- Are there any things that might decrease the negative impact?
- How important is this possible impact? Would you say it is very important, somewhat important or not very important?

14. How might the benefits and risks of biosolids compare to other fertilizers that may be used in your area (*referenced in Q. 11*)?

Interviewer read: We are interested in knowing the thoughts and questions you may have about germs, bacteria and other living organisms, which are collectively known as pathogens. Examples of pathogens are E.coli and salmonella. Pathogens occur naturally in wastewater and are reduced or removed when sewage sludge is treated to produce biosolids. In order to ensure safe levels of pathogens in biosolids land application, the City of Tulsa follows all government regulations on how biosolids are produced and applied to land.

15. Do any additional thoughts or questions come to mind as I read this description?
    Prompt if not mentioned:
    - What could happen if pathogens were not reduced to safe levels in biosolids land application?
      Prompt if not mentioned:
      - To the land?
      - To water?
      - To animals?
      - To people?
    - Who could be affected?

16. Can you think of other things that could potentially have unsafe levels of pathogens?
    - Can you give an example?
    - Anything specifically on farms?
    - *If not mentioned:* What about other types of fertilizers?
    - How would you compare the risks of biosolids to the risks of these things? Would you say you would generally be more concerned, less concerned, or about the same? Why do you say this? *Interviewer, ask question for each possible farm source and/or fertilizer.*

Part 3: Decision-Making Process for Biosolids Land Application

I'd now like get your thoughts on how decisions about biosolids land application may be made.

17. Who do you think are the key people making decisions about biosolids land application in your county?
    - How do you think they are involved in these decisions? What do you think they are responsible for?
    - Is there anyone else who may influence decisions to apply biosolids to land in your county? How do you think they might influence these decisions?

18. What is the most important thing these key people should consider when making decisions about biosolids land application? {*Interview, prompt for specific things. For example*, what the land is used for?}

19. Who would you trust most to make good decisions about biosolids land application?
    - Is there anyone that you have less trust in?

20. Given what you know now, would you say that you are generally in favor of biosolids land application in your county, generally against biosolids land application your county, or undecided?
    - Can you explain your answer?
    - Are there any things that would make you more likely to be in favor of biosolids in your county?
    - Are there any things that would make you less likely to be in favor of biosolids land application in your county?

Part 4: Perceptions on Communications Related to Biosolids Land Application

Now I would like to get your thoughts on communicating with people about biosolids.

21. Having discussed biosolids and their use in land application, what do you think is the most important information that should be shared with people in communities where biosolids are applied?
    - What are the most important questions that you would like answered? Why?

22. Who would be the best person or organization to share this information with people in your county?
    - Why?

23. What would be the best way for City of Tulsa Public Works Department to communicate with you and others in your county about biosolids? {*If an example is needed, say* "What about through a newsletter mailed to you?"}
    - Why would this be useful for you?
    - Are there any other ways?

Part 5: Closing

We are near the end of the interview.

24. Do you have any other comments or questions that we have not yet addressed?

25. If you could offer the organizations that produce or use biosolids one piece of advice, what would that be? Why would that be important?

Demographics

Those are the main questions that I had. In closing, I have a few demographic questions.

26. Which age group are you in: 20s, 30s, 40s, 50s, or 60 or more?

27. How long have you lived in the area?

28. Do you own or rent your residence?

29. Do you have children under 18 living at home?

30. What is your occupation?

31. What is your annual household income: less than $25,000, $25,000-$50,000, $50,000-$75,000; $75,000-$100,000; $100,000-$125,000 or more than $125,000?

32. [Interviewers, record gender]

Those are all the questions that I had. You've been very helpful. Thank you for your time. You've raised some important questions in the course of this interview. For more information on biosolids land application around Tulsa, you may call Cindy Crider at 918-596-9453. Or, if you would like to give us your permission, we can pass your name and telephone number along to the Public Works Department to follow-up with you.

If they have <u>questions about the research</u>: The results will be used by WERF to improve communications with communities about biosolids and their use in land application. If people have questions about the research or WERF, forward them to dprc@decisionpartners.com or 877-588-9106 (or take their number and DP will contact them).

We will offer you $30 in appreciation for your participation in this interview.

Turn off recorder: May I please have your complete mailing address so that we can forward the $30.00 honorarium?

APPENDIX F

SUPPLEMENTARY WORKSHEETS TO SUPPORT USE OF THE PRIMER FOR BIOSOLIDS PROFESSIONALS

F.1 Introduction

A key deliverable of this Research Challenge is the *Primer for Biosolids Professionals: Conducting Effective Community Outreach and Dialogue on Biosolids Land Application* (referred to as the Primer in this report). See Chapter 4.0. The Primer was developed as stand-alone guidance document to supplement this Research Challenge Report. It provides the principles of risk communications and guidance on the tasks and tools to use in each step of the process.

Included in the Primer are worksheets that biosolids professionals can use to complete the step. In many cases, the sample worksheets in the Primer are partially completed, drawing on the experience and learnings gained from the Research Challenge.

This appendix contains the full set of blank worksheets, without the sample completions. Biosolids Communications Teams can use pull these worksheets out to customize and use for their own purposes.

Note: These worksheets should be used as part of the Strategic Risk Communications Process described in the Primer, and with close attention paid to the instructions, tips, guidance and samples provided in the Primer for each worksheet.

Worksheet 1-1. Form the Biosolids Communications Team.

Planning Team Name:	
Team Member Name, Title	**Roles and Responsibilities**
Core Planning Team	
	Sponsor • Ensures adequate staffing and resources. • Promote communications opportunity to senior leadership. • Other duties:
	Team Lead • Manages team process to achieve the Opportunity defined by the Team, supported by the Sponsor. • Ensures quality control. • Reports progress regularly to the Sponsor. • Other duties:
	Project Support • Manages team communications, sets agendas etc. • Documents meetings and processes. • Supports in consultation and research activities. • Other duties:
	Other Team Members • Support the Team Lead in the process of developing, implementing and evaluating outreach and dialogue. • Other duties:
Technical or Communications Support (attends planning meetings as requested)	
	Technical Expert • Provides in-depth technical support to the team. • May support the team by making presentations and responding to questions of technical nature. • Other duties:

Planning Team Name:	
Team Member Name, Title	**Roles and Responsibilities**
	Regulatory Expert • Responds to questions of regulatory nature. • Delivers presentations focused on regulations. • Other duties:
	Web master • Ensures website function. • Collects, organizes, and compiles responses to public comments. • Other duties:
Key Stakeholders* (to be advised/involved as needed or as requested)	
	List:
* These stakeholders may include landowners, Department of Health Officials, State Health Officials, etc.	

Worksheet 1-2. Opportunity Statement.

OPPORTUNITY STATEMENT
State the opportunity as a sentence: The opportunity of the <<name>> Team now is to << do what>> <<for what purpose?>> We will do this <<how?>> <<by when?>> The opportunity of the ______________________ Team now is to: because: We will accomplish this by: We will measure our success by:

Worksheet 2-1. Guiding Questions for a Technical Assessment.

Guiding Questions for a Technical Assessment

1. How are biosolids currently managed as part of your biosolids land application program? Consider:
   a. Who are the key players and what role do they have?
   b. What are the key features of the treatment process?
   c. What is the size and scope of your biosolids program operations?
   d. What are the types of biosolids uses in your community(ies)? Include land application and other uses.
   e. Do you dispose of biosolids (e.g., landfill)? Under what conditions?
   f. *Compile a list of references, such as technical documentation.*

2. Describe the community where you operate land application program. For each community:
   a. How would you define the type of community (e.g., primarily rural, agricultural, heavy tourism, mixed)?
   b. Describe any particular features of the community -- such as geography, topography, climate -- that should be taken into consideration when or communicating about your biosolids land application program.
   c. What type of people live in this community (e.g., people moving in from the city? ranchers? farmers? people with young families?)

3. Describe a typical land application in this community: (e.g., forest application? small farms or large farming operations? What types of crops? In more rural areas or near residential neighborhoods?).

4. Have you conducted biosolids land application in this community(ies) previously? If so, when? What was the outcome? Were there any community issues that had to be resolved? If so, how were they resolved?

5. Describe the key regulations related to biosolids management in the community(ies) where you operate. Consider federal, state and local. Are there any particular issues or challenges your program faces to follow these regulations?
   a. Describe the restrictions, (e.g., how far are the buffers?). *Compile a list of references.*
   b. Describe the regulatory oversight (e.g., county monitors). *Compile a list of references.*

6. What are the key benefits of the land application of biosolids in these communities? Consider environmental, economical, societal and other benefits. For each: Who benefits? Why is this important? *Compile a list of references.*

7. What are the potential safety risks (e.g., pathogen contamination, health impacts due to odor, etc.) of the proposed biosolids land application? (Consider things such as possible risks associated with severe weather events and biosolids management failures). For each:
   a. What is currently known about this risk? What is still uncertain about this risk? What is currently being done to better understand these issues?
   b. What technical resources, such as reports and risk assessment tools are available? *Compile a list of references.*
   c. What is currently being done to address or manage this risk in your community(ies)?

8. What are the key potential environmental risks of biosolids land application in this community? (Consider things such as possible risks associated with severe weather events and biosolids management failures). *Address the detailed questions in Q6.*

9. Has odor been an issue in this community? If yes, when was it and how significant was it?
   a. Describe any issues related to odor: Where?
   b. How do you know? Have there been complaints by near neighbors? Others? Why?
   c. What is currently being done to address or manage odor issues? What more could be done?

10. Describe any significant management issues, such as incidents where biosolids were inappropriately managed or negative impacts occurred as a result of biosolids management: What happened? Who was involved? How was it resolved? What impact did it have on community stakeholders' judgment?

11. Describe your current communications processes in this community (e.g., notification process, questions directed to whom) or materials (e.g., pamphlets)?
    a. What communications materials have been used in this community? Compile a sample.
    b. Are there any communications sources that you rely on to communicate currently with your stakeholders (e.g., send people to a particular website?)

Worksheet 2-2. Stakeholder Map.

STAKEHOLDER MAP

Insert the relevant stakeholders/names in the appropriate circle based on the definitions in the Primer.

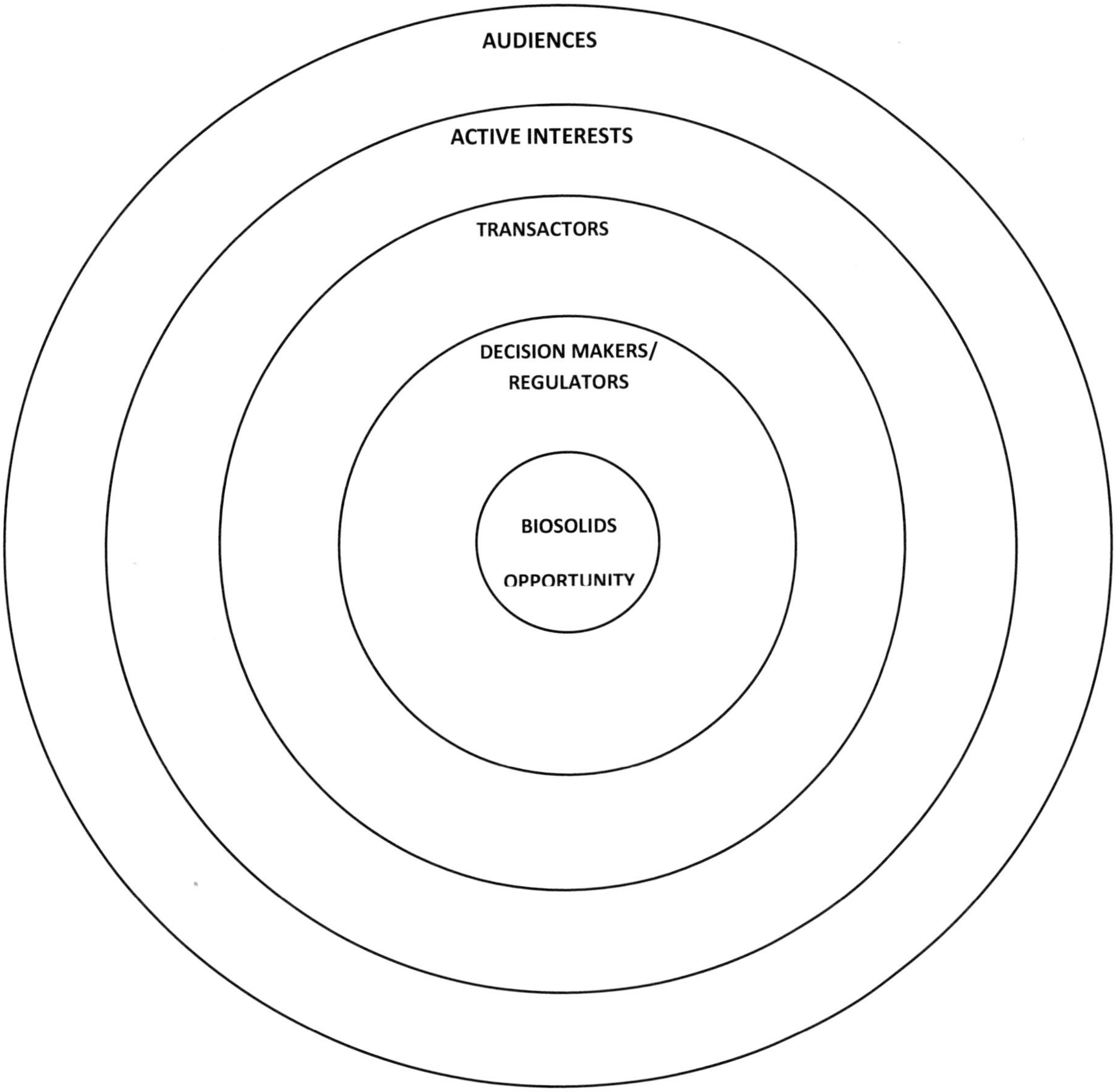

Worksheet 2-3. Key Stakeholders for Community Dialogue.

KEY STAKEHOLDERS FOR COMMUNITY DIALOGUE		
Stakeholder	Type of Stakeholder	Why Priority?

Worksheet 2-4. Stakeholder Hypothesis Development.

STAKEHOLDER HYPOTHESIS Key Stakeholder: ____________________________________	
Guiding Questions	**Priority Stakeholder Hypothesis**
What do they know now about the biosolids topic/risk that is correct? How do we know?	
What don't they know or misunderstand that is consequential? How do we know?	
What might they want to know? How do we know?	

STAKEHOLDER HYPOTHESIS **Key Stakeholder:** ______________________	
Guiding Questions	**Priority Stakeholder Hypothesis**
Who do they trust, and why? How do we know?	
What communication methods do they prefer? How do we know?	
What do they value related to biosolids risk management? How do we know?	
What are the key factors that they consider when making decisions? How do we know?	

Worksheet 3-1. Stakeholder Research Outline.

Stakeholder Research Outline
Use this outline as a foundation to develop your stakeholder research interview protocol or questionnaire. Questions should be designed to learn individual stakeholders': • *Familiarity with the term 'biosolids'. May provide a simple definition after hearing first thoughts.* • *Anything seen, heard or learned about biosolids and through what channels.* • *Reasons why biosolids may be produced and applied to land.* • *Biosolids production and land application processes and who is involved.* • *Potential positive and negative impacts associated with biosolids land application. May want to ask prompts for specific impacts, like agricultural benefits, human health risks, odor.* • *Comparison of the benefits and risks of biosolids to those of other types of fertilizers.* • *Understanding of decision makers and other 'key players' in biosolids land application biosolids projects.* • *Understanding of government regulations on biosolids land application. May want to ask about people's thinking on specific regulations.* • *Priorities related to biosolids management.* • *Overall judgments of local land application projects.* • *Questions that should be answered about biosolids.* • *Preferred and trusted communication sources and methods.*

Worksheet 3-2. Research Analysis.

Key Stakeholder: ______________________________		
Guiding Questions	**What did we learn from Stakeholders?**	**How does this compare to our hypothesis?**
What do they know now about the biosolids land application that is correct? *Use the Technical Assessment as a guide.*		
What don't they know or misunderstand that is consequential? *Use the Technical Assessment as a guide.*		
What do they want to know?		

Key Stakeholder: ____________________		
Guiding Questions	**What did we learn from Stakeholders?**	**How does this compare to our hypothesis?**
Who do they trust, and why?		
What communication methods do they prefer?		
What do they value most about biosolids land application?		
What are the key factors that they consider when making judgments and decisions about biosolids land application?		

Worksheet 4-1. Generic Outreach and Dialogue Plan.

DATE	**ACTION**	**WHO**

Worksheet 4-2. Biosolids Dialogue Pretests Facilitators' Guide.

Biosolids Dialogue Pretests Facilitators' Guide

One-on-one discussion with each participant:

1. What are your first impressions of this presentation?
2. What were some of the things that you took away from this presentation? *Probe for specifics.*
3. Did any new thoughts about biosolids come to mind after listening to this presentation?
4. What [*other*] questions came to mind? *Probe for specific questions. Note them, but do not provide any answers at this time. We will address those in the detailed discussion.*
5. What in the presentation did you find most useful? Why?
6. What in the presentation did you find least useful? Why?
7. What, if anything, in the presentation raised concerns in your mind?
8. Are there any topics that we didn't cover, or that you think we should cover in more detail?
9. How useful do think that a dialogue session with a presentation like this one today will be to community members? Why or why not?
10. How else might we improve the presentation? Why would this be an improvement?
11. Do you have any other comments or questions on the presentation or about biosolids land application more generally? *Probe for specifics, as time permits. Use back of sheet for notetaking.*

Hand the person a brochure to read in advance of the group discussion. Invite them to get some more refreshments

Group Discussion

1. *Opening on presentation.* Thank you for offering your detailed feedback on the **presentation**. Can we ask each of you to share 1 or 2 things that you though were most important or interesting about the presentation?
2. What are your first impressions of the **brochure?**
   - What were some of the things that you took away from brochure?
   - Which of these messages is the most important, do you think?
   - Did it raise any other questions in your mind?
   - Are there any topics that we didn't cover, or that you think we should cover in more detail?
   - How useful do think that a pamphlet like this one today will be to community members? Why do you say this?
3. There are a few specific topics raised in the presentation and the brochure that we'd really like to get your thoughts on:
   - What about the idea that the City is committed to **engaging with** landowners, neighbors and other community members.
     i. What comes to mind when you hear that?
     ii. Does it raise any questions?
   - What about the idea that biosolids are a **recycled product?** *Use similar prompts as in Q3.*
   - What about the **benefits of biosolids to the community?**
   - What about the idea that **biosolids are safe**?
   - Time permitting (and if not mentioned),
     i. Odor

ii. Environmental benefits – for, example, trapping CO_2 in the environment.
iii. Others?

4. Regarding the **proposed signage** that would be posted a few days before the land application:
   - How would this be valuable to neighbors in the community?
   - *Landowners:* Does it raise any specific thoughts to you as a landowner that will have the sign on your land?
   - What were your first impressions of the signage?
   - Did it raise any other questions in your mind?
   - Is there any other information you'd like on this sign?
   - How else might we improve the sign?
   - *Show two different versions:* We have some other examples of signs. Any comments? Do you have a preference for one? Why do you say that?

5. *Conclusion Question – High Priority.*
   - How do you think that these proposed dialogue meetings, and all of the communication materials will help us better engage with community members? Why would this be helpful?
   - What could we do to improve?

6. Were there any questions about biosolids that you have that we have not yet addressed? *Time permitting, address questions.*

7. Are there any other Comments?

Worksheet 5-1. Meeting Evaluation Form.

Feedback Card
Meeting Location: ______________________________ **Date:** ____________________

Please take a moment to share your thoughts on this dialogue session so that we can continue to learn about your interests and priorities, and to improve how future events of this kind are conducted.

1. How did you hear about this dialogue session?
____

2. Please tell us a little about why you decided to attend today?
____
____

3. In general, did the session meet your expectations? Why/why not?
____
____

4. Please provide us with some feedback on our presentation. How well did the information provided address your needs?

☐ Very Well ☐ Well ☐ Somewhat Well ☐ Not at all Well
Please explain your rating:
____

5. How well did you feel that you were you able to ask questions and share your thoughts following the presentation?

☐ Very Well ☐ Well ☐ Somewhat Well ☐ Not at all Well
Please explain your rating:
____

6. Do you have any additional questions or thoughts to share?
____
____

7. How might we improve future consultation sessions of this kind?
____
____

Thanks for taking the time to share your thoughts. If you have any unanswered questions; or would like a team member to get in touch with you, please leave your contact information below:

Your name:
Organization (if applicable):
Address:
Telephone number:
Email address:
Preferred method of contact:

Worksheet 6-1. Checklist for Designing Appropriate Performance Measures for Each Step of the Process.

Process Step/Measure	Comments/Observations
Step 1: Define the Opportunity ❑ Is the Opportunity Statement clear, achievable and measureable? ❑ Are the roles of the Team members clearly defined? ❑ Are adequate resources in place to achieve the Opportunity?	
Step 2: Characterizing the Situation ❑ Has the Technical Assessment been completed and documented? ❑ Are technical resources available to support the Team as required? ❑ Have all key stakeholders been identified? ❑ Have the stakeholder hypotheses been documented? ❑ Is the Team prepared to add to its Stakeholder Map as you learn more through the process? ❑ Has the Team defined its priority stakeholders?	
Step 3: Assess Stakeholders' Interests, Priorities, and Communications Needs ❑ Has the Team conducted research – formal and informal – and identified stakeholder interests and priorities? ❑ Has the Team defined topics likely to be of interest to stakeholders? Is the Team prepared to address these? ❑ Has the Team defined key gaps and alignments between the thinking of the Team and the stakeholders? ❑ Has the Team defined optimal dialogue and information channels based on stakeholder needs and preferences?	
Step 4: Develop and Pre-Test Communications Plan and Materials ❑ Has the Team developed messages and materials aligned with stakeholder interests and priorities? ❑ Has the Team pre-tested these materials to ensure they will be understood as intended?	
Step 5: Developing the Plan ❑ Have you determined the most appropriate dialogue techniques with which to engage stakeholders? ❑ Have the Team considered if your outreach and dialogue efforts should be linked to other existing or upcoming processes either inside	

or outside the organization? ❑ Is there a need to accommodate special relationships within the process? ❑ Are your plans culturally sensitive and appropriate? ❑ Have you developed, pre-tested and refined communication messages? ❑ Do you need a neutral third party to help implement and evaluate outreach efforts? ❑ Have you identified the skills and training that Team members need to conduct effective dialogue? ❑ Have you identified information to support dialogue efforts? ❑ Is the information in a format and at a level that stakeholders can understand? ❑ Do you plan to collect feedback throughout the process? ❑ Do you have a plan for how internal communications on the process be managed? ❑ Do you have a media relations plan? ❑ Do you have a plan to communicate other key stakeholders about the process and its outcomes? ❑ Is your consultation, outreach and dialogue plan consistent with your organization's communication policy?	
Step 6: Implement the Communications Plan ❑ Have the logistics to support your plans – invitations, meeting facilities, etc. – been carefully planned? ❑ Does the Team have the resources it needs? ❑ Do Team members have the necessary dialogue skills? ❑ Are Team members and other participants clear on their roles and responsibilities?	
Step 7: Measure Process and Outcomes ❑ Did the team achieve its Opportunity? ❑ Did the process enable meaningful participation by key stakeholders in their judgment? ❑ How might the process be improved in the future? ❑ What were the Team's key learnings? What did the Team do well? What could have been improved?	

APPENDIX G

STRATEGIC RISK COMMUNICATIONS WEF RESIDUALS AND BIOSOLIDS 2010

Workshop Agenda

May 23, 2010

8:00-8:10 am	Welcome and Introductions (Alan Hais), Overview of the Workshop (Patrick Gurian)
8:10-8:30 am	Integrated Risk Assessment and Communications Approach/Tools for Stakeholder Engagement (Sara Eggers)
8:30-8:45 am	Microbial Risk Assessment Tools for Addressing Scenarios of Concern (Patrick Gurian)
8:45-9:05 am	Case Study 1: Tulsa (Sara Eggers)
9:05-9:25 am	Case Study 2: Virginia (Chris Peot)
9:25-9:45 am	Case Study 3: Pennsylvania (Diane Garvey)
9:45-9:50 am	Charge to Breakout Groups
9:50-10:00 am	Group Assignments and Introductions
10:00-10:15 am	Break
10:15-11:00 am	Breakout Group Work Time
11:00-11:50 am	Breakout Groups Report and Panel Discussion (Panel: Beecher, Eggers, Garvey, Meckes, Peot, Pepper, Wardell, facilitated by Gurian)
11:50-12:00 pm	Wrap up: Recommendations and Outstanding Needs (Gurian and Hais)
12:00-1:00 pm	Optional lunch: Revisions to Scenario Plans, Review of Workshop

Workshop Agenda

Assessing Risks and Effective Communication of Land Applied Biosolids
September 14, 2010
Village Inn – Clemmons, NC

Purpose: This seminar will consist of how to assess and communicate the risks of land applied residual biosolids. A morning workshop will include hands-on training and working in break-out groups to develop proper communication techniques given different scenarios. Also during this seminar you will get an opportunity to get hands on training of a Microbial Risk Assessment Tool, learn about new government lobbying efforts for wastewater in NC, and hear about proven marketing techniques for biosolids programs.

Who Should Attend: Biosolids Mangers, Regulators, Soil Scientists, Agronomists, Engineers, Researchers

Contact Hours/PDH's

- **WASTEWATER OPERATORS:** *This seminar is approved by the WPCSOCC for 6.0 contact hours for continuing education. WPCSOCC Course ID: CE06231004*
- **PROFESSIONAL ENGINEERS:** *This seminar is approved for 6.0 PDH's.*

Direct Questions Related to Technical Content to: Scott Carpenter, *Soil Plus LLC*, 252.917.4288

Registration Fees: *(includes lunch)*
Pre-Registration: Members $105, Non-Members $130
On-site Registration: Members $125, Non-Members $150
Students: $25

Agenda:

8:00 am	**Registration, Pre-workshop Survey**
8:30 am	**Integrated Risk Assessment and Communications Approach/Tools for Stakeholder Engagement** – *Sara Eggers (Decision Partners)*
8:50 am	**Microbial Risk Assessment Tools for Addressing Scenarios of Concern** – *Patrick Gurian (Drexel University)*
9:05 am	**Case Study 1: Tulsa** – *Sara Eggers (Decision Partners)*
9:25 am	**Case Study 2: Virginia** – *Robert Crockett (Virginia Biosolids Council)*
9:45 am	**Case Study 3: Pennsylvania** – *Michael Wardell (Mid-Atlantic Biosolids Association)*
10:05 am	**Charge Break-out Groups/Group Assignments and Introductions**
10:15 am	**Break**
10:30 am	**Break-out Groups Work Time**
11:15 am	**Break-out Groups Report and Panel Discussion** – *Ned Beecher, Michael Wardell, Patrick Gurian, Chris Peot, Diane Garvey*
12:00 pm	**Lunch**
1:00 pm	**Hands on training Using a Microbial Model** – *Patrick Gurian (Drexel University)*
2:30 pm	**Break**
2:45 pm	**Proven Marketing Techniques for Biosolids Programs** – *Michael Wardell (Mid-Atlantic Biosolids Association)*
3:20 pm	**NC Water Quality Association**
4:00 pm	**Adjourn & Distribution of Certificates of Attendance**

Assessing and Communicating Risks from Pathogens in Land Applied Biosolids

Strategic Risk Communications

Sara Eggers, Decision Partners

WEF *Residuals and Biosolids* 2010
May 23, 2010

WERF Challenge SRSK2R08

Bottom Line Up Front

- Stakeholder judgment and decision making are complex.
- Strategic Risk Communications is integral to effective risk management.
- Strategic Risk Communications is supported by robust processes, methods and tools.

WERF Challenge SRSK2R08

Who are Stakeholders?

- Stakeholder: Any individual, group, or organization that may affect, be affected by, or perceive itself to be affected by a decision:
  - Municipal wastewater treatment facilities, who are mandated to safely manage wastewater by-products.
  - Biosolids companies, who take advantage of the business opportunities.
  - Landowners, who find agricultural and economic value in biosolids land applications.
  - Neighbors, who do not directly benefit from biosolids land application (but who may affected in other ways).
  - Elected officials who weigh the tradeoffs of biosolids applied in their communities and make decisions about their acceptability.

WERF Challenge SRSK2R08

The Nature of Stakeholder Judgment

- Individual stakeholders judge the benefits, risks and tradeoffs of an issue or opportunity based on:
  - Perceived facts.
  - Perceived options (and their benefits and risks).
  - Perceived match with values.

- Typical core issues influencing stakeholders' judgments:
  - Acceptability of risk.
  - Trust in decision makers and communicators.
  - Control over choices.
  - Fairness and respect.
  - Quality of communications and transparency of action.

WERF Challenge SRSK2R08

The Nature of Stakeholder Judgment

Influences on Acceptability of Risks

Less Acceptable	*More Acceptable*
Low trust	High trust
Unclear benefits	Benefits understood
Not controllable	Individual control
Involuntary exposure	Voluntary exposure
No alternatives	Alternatives available
Unfair distribution	Fair distribution
Dreaded consequence	Common consequence
Affects children	Affects everyone
Human origin	Natural origin
High media concern	Low media concern
High symbolism	Low symbolism

▪Source: Slovic, Fischhoff et al

WERF Challenge SRSK2R00

The Nature of Stakeholder Judgment

Influences on Judgments

- People's judgments and decision making are influenced primarily by communications:
  - Communications affect how people perceive facts, options, and their tradeoffs.
  - Communications include what people read, hear and see.
- There are no universally acceptable solutions. A solution that is "right" for one stakeholder, or one community may not be "right" in another.

WERF Challenge SRSK2R00

Beliefs that Can Hinder Communication Efforts

- Laypeople are ignorant and irrational about risk.
- Science can resolve all issues.
- It is easy to predict people's preferences, beliefs and how they will interpret information:
  - Knowing what people believe is the same as knowing why they believe it.
  - "To believe what I believe, people must know what I know."
- Information out = communication.
- Communication will increase liability.
- Communication will lead to a loss of control.

WERF Challenge SRSK2R06

Conventional, One-Way
"Tell" Process

"Experts" identify the issues and goals

↓

Select "audiences"

↓

Decide what audiences need to know

↓

Develop key concepts and messages

↓

Produce messages to fit management mindsets

↓

Publish and present

↓

Determine success, without evaluation

WERF Challenge SRSK2R06

Conventional Strategies Can Fail

- Traditional approaches to communication – like "Tell" "Decide-and-Defend" and "Get the message out" – can be surprisingly ad hoc.
- Such strategies can often underperform or fail because:
  - The issues and opportunities are complicated.
  - Multiple stakeholders are involved.
  - The science and facts are not available or are controversial.
  - Benefits and risks may be uncertain or unknowable.
  - The communication environment is sensitive and complex.
  - Communications efforts are not integrated with risk management.
  - The credibility of leaders may be under intense pressure.

WERF Challenge SRSK2R00

Dialogue-Based

Strategic Risk Communications

- Strategic Risk Communications in an action-oriented, purposeful process of skilful interaction supported by relevant information.
  - Integral to risk management and other decision-making frameworks.
  - Key Goal: To enable decision makers and stakeholders to make well-informed and appropriate decisions.
- Strategic Risk Communications draws from multiple disciplines:
  - Risk management, decision science, engineering.
  - Risk perception, risk communication, behavioral decision research, psychology.

WERF Challenge SRSK2R00

Strategic Risk Communications

Guiding Principles

- Create a effective collaboration among technical, management and communications professionals.
- Learn what people already believe and why they believe it.
- Tailor communications to what people need to – and want to – know about the decisions they face.
- Pre-test communications strategies and messages to ensure effectiveness.
- Be sensitive to behavior as communication.
- Measure communications processes, effects and outcomes for continuous improvement.

WERF Challenge SRSK2R08

Dialogue-based

Strategic Risk Communications Process

Define the Opportunity

Characterize the Situation

Assess Stakeholder Perceptions of Risks, Benefits & Tradeoffs

Assess How Stakeholders Perceive the Options

Develop & Pre-test Strategies, Risk Communications Plans & Messages

Implement Risk Communications Plans

Evaluate Risk Communications Effectiveness

WERF Challenge SRSK2R08

Strategic Risk Communications is Supported by Robust Tools

For example:

- Stakeholder Hypothesis: Tools to identify stakeholders, and integrate current understanding about their interests, beliefs and values.
- Expert Modeling: Analytical method to summarize and integrating expert understanding on complex risk issues and related decisions.
- Mental Model Research: Science-led method to discover people's values, interests, priorities, beliefs and informational needs.
- Communication Plans: Tools and guidance to translate insight into sound communications plans, processes and messages.
- And others tools adapted as needed.

WERF Challenge SRSK2R00

SAMPLE Expert Model of Impacts of Biosolids Land Applications

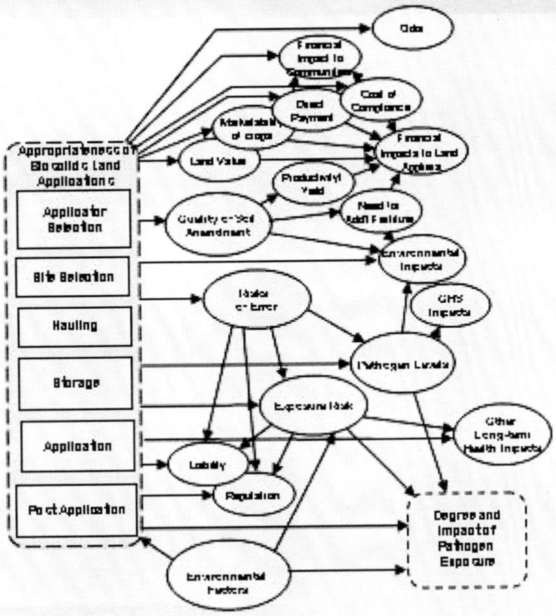

WERF Biosolids Risk Assessment and Risk Communications (SRSK2R03)

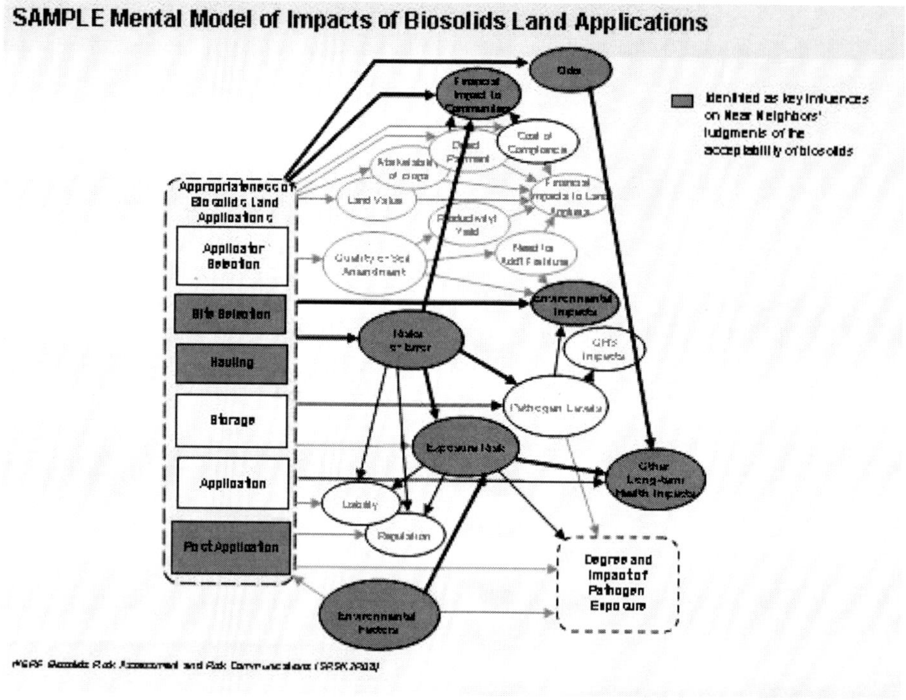

Strategic Risk Communications

Process Benefits

- The Strategic Risk Communications Process can produce good results because:
  - It enables decision making.
  - It is strategic (emphases planning).
  - It is robust (driven by insight and research).
  - It is adaptable to complex situations.
  - It is respectful.
  - It enables continuous improvement.

WERF Challenge SRSK2R08

Assessing and Communicating Risks from Pathogens in Land Applied Biosolids

Tulsa Case Study

Sara Eggers, Decision Partners

May 23, 2010

WEF *Residuals and Biosolids* 2010

Savannah, Georgia

WERF Challenge SRSK2R08

Tulsa Case Study Objectives

- Decision Partners is leading the risk communications component of the WERF Research Challenge: Biosolids Pathogens Risk Assessment and Risk Communications.
- DP is adapting its Strategic Risk Communications Process to accomplish the work, though two case studies.
- Case study objectives:
  - Refine and validate the approach applied to biosolids land applications.
  - Inform the development of prototypical strategies that can be adapted by utilities and other industry stakeholders.
  - Inform development of the pathogen risk analyses and tool by the Research Challenge's Biosolids Risk Assessment Team.

WERF Challenge SRSK2R08

Strategic Risk Communications Process

Define the Opportunity

Characterize the Situation

Assess Stakeholder Perceptions of Risks, Benefits & Tradeoffs

Assess How Stakeholders Perceive the Options

Develop & Pre-test Strategies, Risk Communications Plans & Messages

Implement Risk Communications Plans

Evaluate Risk Communications Effectiveness

WERF Challenge SRSK2R00

Tulsa Biosolids Operations Overview

- The Tulsa Case Study may be considered "business as usual":
  - Biosolids operations began in 1984.
  - ~ 9,000 dry tons applied annually.
  - Biosolids are produced by mesophilic anaerobic digestion and are land applied as either liquid or cake (16% water).
  - 8 full-time staff, 2 contractor hauling companies.
  - Contributed industry best practices.
  - Demand for biosolids land applications outstrips supply.
  - Very limited vocal opposition to biosolids land applications.
  - No significant formal communications processes.

WERF Challenge SRSK2R00

Tulsa Biosolids Team

- City of Tulsa Public Works
  - Tom Krueger, Manager, Water Pollution Control
  - Brian Bolte, Support Services Manager
  - Cindy Crider, Administrative Operations Supervisor
- WERF Project Team
  - Sarah Thorne, Decision Partners
  - Sara Eggers, Decision Partners
  - Alan Hais, WERF
  - Patrick Gurian, BRAT

WERF Challenge SRSK2R08

Tulsa Case Study

Overview of Project Tasks

Process Steps	Tasks	Dates
Steps 1 & 2 Tulsa Team Workshop	• Developed Opportunity Statement • Characterized the Situation: • Expert Model • Stakeholder Hypotheses	May 2009
Steps 3 & 4 Assess Stakeholder Perceptions	• Conducted 24 interviews (Landowners & Neighbors) • Coded and analyzed interview data • Shared report and recommendations with Team	July – Nov. 2009
Steps 5 & 6 Develop and Pretest Communications Plan and Materials	• Developed process, brochure and presentation • Hosted pre-test community dialogue sessions • Refined Plan and materials	Dec 2009 – April 2010

WERF Challenge SRSK2R08

Step 1 Results

Tulsa Biosolids Opportunity

- The Opportunity is to ensure long-term sustainability of our business and the societal value it provides by systematically building key stakeholder support for, and effective management of, our current and future biosolids products and operations.
- Our work focuses on applying the Strategic Risk Communications Process in order to:
  - Discover key stakeholder perceptions, interests and priorities related to our products and operations.
  - Proactively engage key stakeholders in dialogue about the benefits and risks of our current and future products and operations.
  - Effectively address stakeholders' interests and priorities through dialogue, supported by relevant information.

WERF Challenge SRSK2R08

Step 2 Results

Tulsa Biosolids Team: Stakeholder Map of Sustainable Biosolids Business

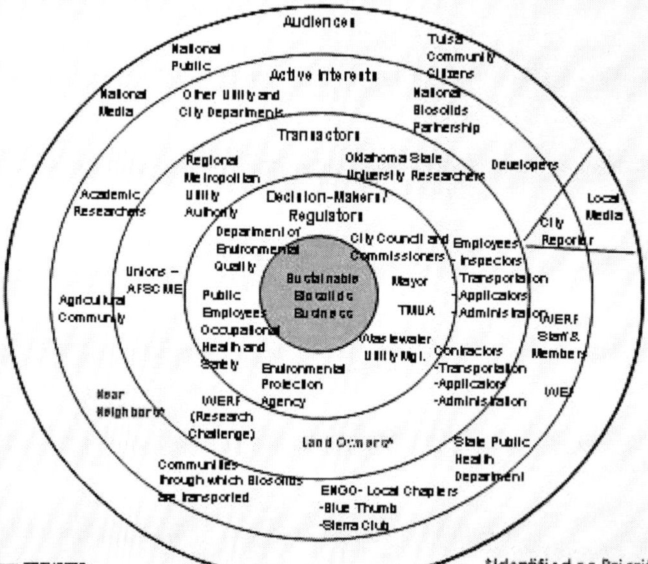

WERF Challenge SRSK2R08

*Identified as Priority Stakeholders for the purpose of the Case Study

Step 3 Results

Top-Line Research Findings

- Interviewees were generally familiar with the biosolids and biosolids land applications.
- Landowners cited direct agricultural and economic impacts.
  - Neighbors talked about indirect positive impacts – wastewater management and benefits to their neighbors.
- Interviewees talked most about odor and safety.
  - Neighbors: odor is persistent, significant and has negative health and quality of life impacts.
  - Landowners: odor is "minor", "short-lived" and "worth it".
  - Other negative impacts: hazardous waste, runoff, over-application, contamination, animal health, and uncertainty.

WERF Challenge SRSK2R08

Step 3 Results

Top-Line Research Findings *continued*

- Most important considerations in decisions: quality and oversight of regulations, safety, and the impact on neighbors.
- Landowners were in favor of biosolids land applications. Neighbors were generally not.
- – Landowners: trust in the regulations and the City, and the "naturalness" of biosolids.
- – Neighbors: lack of trust in decision makers, lack of confidence in the regulations and the uncertainty of biosolids.
- Opportunities to improve communications:
- – Provide more transparency in decision making.
- – Communicate in advance.
- – Provide information on the benefits and risks.

WERF Challenge SRSK2R08

Step 5 Results

Communications Plan

- Developed a draft Communications Plan:
  - Frames biosolids land applications as a beneficial product.
  - Includes a process for communicating with community stakeholders about specific biosolids land applications.
  - Includes plan for pre-testing draft communications materials.
- Drafted standard communications materials:
  - Generic presentation and pamphlet to support communications about biosolids land applications with community members.
  - On-site signage.
- Pretested on March 9 with two small sessions of research participants (Landowners and Neighbors) in Glenpool, OK.

WERF Challenge SRSK2R08

Tulsa

Beneficial Use of Biosolids

This field is benefiting from biosolids produced and land applied by the City of Tulsa.

Approximate Date of Application:
March 31, 2010

Please stay off this field while the sign is posted.

For more information about the City of Tulsa Biosolids Program, please visit our website at www.cityoftulsa.org/biosolids or contact us at (918) 596-9453.

WERF Research Challenge SRSK2R08

Step 5 Results

Communication Dialogue Pretest

Brian Bolte delivering the presentation.

Landowners providing feedback.

WERF Challenge SRSK2R05

Final Steps

- **Refine the Communications Plan and standard materials.**
- **Develop specific plan for engaging with Near Neighbors:**
  - **A process to learn and address their specific questions and priorities.**
- **Develop risk communications specifically for Landowners:**
  - **Clear descriptions of Landowners' role and responsibilities related to biosolids land applications.**
  - **Guidelines on how to effectively communicate and engage with Neighbors.**
- **Document the process and results as part of the WERF project.**

WERF Challenge SRSK2R05

Team Observations on Strategic Risk Communication Process

- Tulsa "would rather be proactive than end up with a crisis down the road".
- Tulsa learned that Landowners and Near Neighbors want to engage in dialogue.
- The Process enables the utility to:
  - Gain insight into who their stakeholders are and what their specific issues and priorities are;
  - Understand where and at what level to initiate communications;
  - Better engage its key stakeholders on their issues and priorities.
- The Process can be adapted for other risk management and risk communications issues in the future.

WERF Challenge SRSK2R08

Acknowledgements

- Alan Hais and his team at WERF for support on the Research Challenge.
- Patrick Gurian and is Biosolids Risk Assessment Team for collaborating on the Research Challenge.
- Tom Krueger, Brian Bolte and Cindy Crider at the City of Tulsa for support and collaboration on the Case Study.
- WERF's Issue Area Team and the Science Research Advisory Panel for advising and reviewing the Research Challenge.
- Sarah Thorne, Gordon Butte and our Decision Partners Research and Support Team.
- Ned Beecher and my other WEF Workshop collaborators.

WERF Challenge SRSK2R08

REFERENCES

Atman, C.J., Bostrom, A., Fischhoff, B., and Morgan, M.G. 1994. Designing Risk Communications: Completing and Correcting Mental Models of Hazardous Processes, Part 1. *Risk Analysis,* 14(5): 779-788.

Beecher, N., Connell, B., Epstein, E., Filtz, J., Goldstein, N., and Lono, M. 2004. Public Perception of Biosolids Recycling: Developing Public Participation and Earning Trust. Water Environment Research Foundation (00-PUM-5), Alexandria, VA.

Beecher, N., Harrison, E., Goldstein, N., McDaniel, M., Field, P., and Susskind, L. 2005. Risk Perception, Risk Communication, and Stakeholder Involvement for Biosolids Management and Research. *Journal of Environmental Quality,* 34: 122-128.

Bostrom, A., Fischhoff, B., and Morgan, M.G. 1992. Characterizing mental models of hazardous processes. *Journal of Social Issues,* 48(4): 85-100.

Butte, G. and Thorne, S. 2005. Opening the Dialogue on Electricity Supply Mix: Stakeholder Consultation Report. Ontario Power Authority.

CAN / CSA – Q850-97 (R2009) Risk Management: Guideline for Decision-Makers. Canadian Standards Organization.

CAN / CSA ISO 31000 Risk Management – Principles and Guidelines (February 2010).

Carpenter, D. 2009. Deconstructing NIMBY: Understanding the Causes of and Solutions to Social Friction. Presentation to the Fellow Program at the Weatherhead Center for International Affairs, Harvard, University, Cambridge, MA.

Darisi, T., Thorne, S., and Iacobelli, C. 2005. Influences on Decision Making for Undergoing Plastic Surgery: A Mental Models and Quantitative Assessment. *Plastic and Reconstructive Surgery,* 116: 907-916.

Devine-Wright, P. 2005. Beyond NIMBYism: towards an integrated framework for understanding public perceptions of wind energy. *Wind Energy*, 8(2): 125-139.

Downs, J.S., Bruine de Bruin, W., and Fischhoff, B. 2008. Parents' vaccination comprehension and decisions. *Vaccine,* 26: 1595-1607.

Downs, J.S., Bruine de Bruin, W., Fischhoff, B., Hesse, B., and Maibach, E. 2010. How people think about cancer: A mental models approach. In D. O'Hair (ed.), Handbook of risk and crisis communication. Mahwah, NJ: Erlbaum.

Eggers, S.L., Ackerlund, S., Thorne, S.L., and Butte, G. 2010. Growers' Understanding and Implementation of FDA's GAPS Guidance: Mental Models Research Report to the U.S. Food and Drug Administration [Accessed 01/10/11 at http://www.fda.gov/Food/FoodSafety].

Eggers, S.L., Verrill, L., Bryant, C. and Thorne, S.L. (in press). Developing consumer-focused risk communications strategies related to food terrorism. *International Journal of Food Safety, Nutrition and Public Health.*

Fischhoff, B. 1995. Risk Perception and Communication Unplugged-20 Years of Process. *Risk Analysis*, 15: 137-45.

Fischhoff, B., Bostrom, A., and Quadrel, M.J. 1997. Risk perception and communication. In: Detels, R. et al. (eds.) Oxford textbook of public health. London: Oxford University Press, 1997: 987-1002.

Flynn, J., Slovic, P., and Mertz, C.K. 1994. Gender, Race, and Perception of Environmental Health Risks. *Risk Analysis* 14(6): 1101-1108.

Gregory, R., Fischhoff, B., Butte, G. and Thorne, S. 2003. A Multi-Channel Stakeholder Consultation Process for Energy Deregulation. *Energy Policy,* 31:1291-9.

Health Canada. 2006. Strategic Risk Communications Framework for Health Canada and the Public Health Agency of Canada. Minister of Health Canada (ISBN: 0-662-44596-1).

ISO. 2009. ISO 31000:2009 Risk Management – Principles and Guidelines.

Johnson-Laird, P.N. 1983. Mental Models. Cambridge, MA: Harvard University Press.

Johnson, B.B. and Chess, C. 2003. How reassuring are risk comparisons to pollution standards and emission limits? *Risk Analysis*, 23: 999-1007.

Krogmann, U., Gibson, V., and Chess, C. 2001. Land application of sewage sludge: perceptions of New Jersey vegetable farmers. *Waste Management Research,* 19: 115.

Kuhn, R. and Ballard, K. 1998. Canadian Innovations in Siting Hazardous Waste Management Facilities. *Environmental Management,* 22(4): 533–545.

Morgan, M.G., Fischhoff, B., Bostrom, A., and Atman, C.J. 2002. Risk Communication – A Mental Models Approach. New York: Cambridge University Press.

Morgan, M.G., Fischhoff, B., Bostrom, A., Lave, L. and, Atman, C.J. 1992. Communicating Risk to the Public. *Environmental Science & Technology,* 26: 2048-2056.

National Biosolids Partnership. 2005. National Manual of Good Practices for Biosolids. Alexandria VA.

National Research Council. 2002. Biosolids Applied to Land: Advancing Standards and Practices. National Academy Press, Washington, D.C.

North East Biosolids and Residuals Association. 2007. A National Biosolids Regulation, Quality, End Use, and Disposal Survey Final Report [Accessed 1/10/2011 at http://www.nebiosolids.org/uploads/pdf/NtlBiosolidsReport-20July07.pdf].

Palmgren, C., Morgan, M.G., Bruine de Bruin W., and Keith, D. 2004. Initial public perceptions of deep geological and oceanic disposal of carbon dioxide, *Environmental Science and Technology,* 38 (24) 6441–6450.

Parkin, R., Ragain, L., Embrey, M., Peters, C., Butte, G., and Thorne, S. 2004. Risk Communication for Emerging Contaminants, (Project Number 2776) Denver, CO.: American Water Works Association.

Pepper, I.L., Zerzghi, H., Brooks, J.P., and Gerba, C.P. 2008. Sustainability of Land Application of Class B Biosolids. *Journal of Environmental Quality*, 37: 1-10.

Robinson, K.G. and Robinson, C.H. 2006. Biosolids Recycling: An Assessment of Public Perception and Knowledge. Proceedings of the Water Environment Federation, WEFTEC 2006: Session 11 through Session 20, pp. 1070-1077(8).

Roth, E., Morgan, M.G., Fischhoff, B., Lave L., and Bostrom, A. 1990. What Do We Know About Making Risk Comparisons? *Risk Analysis* 10: 375-87.

Santos, S.L. and McCallum, D.B. 1997. Communicating to the public: Using risk comparisons. *Human and Ecological Risk Assessment,* 3: 1197-214.

Shichor, D. 1992. Myths and Realities in Prison Siting. *Crime & Delinquency,* 38: pp.70-87.

Simonson, R. 1992. Managing community opposition to subdivision development. *Real Estate Review*, 22: 82-86.

Slovic, P. 2000. The Perception of Risk. London: Earthscan Publications, Ltd.

Slovic, P., Fischhoff, B., and Lichtenstein, S. 1982. Why study risk perception? *Risk Analysis,* 2: 83-94.

Sjöberg, L. 2004. Local Acceptance of a High-level Nuclear Waste Repository. *Risk Analysis,* 24(3), 737-749.

Takahashi, L. and Dear, M. 1997. The changing dynamics of community opposition to human service facilities. *Journal of the American Planning Association*, 63: 79-93

Thorne, S. and Butte, G. 2007. Deconstructing "NIMBY": Challenging Conventional Wisdom. Annual Canadian Power Conference and Networking Centre, Toronto, ON, November 13.

Thorne, S., Butte, G., Ramsay, L., and Chatigny, E. 2006. Strategic Risk Communications Handbook. Health Canada and the Public Health Agency of Canada.

Thorne, S., Eggers, S.L., Ostrove, N., Buck, E., and Dirolf, K. 2009. Evaluation of How to Communicate to Healthcare Providers about the Risks and Benefits of Prescription Drug Use for Pregnant and Nursing Women. Society for Risk Analysis (SRA), Annual Meeting, Baltimore, MD, December.

U.S. Presidential/Congressional Commission on Risk Assessment and Risk Management Process.1997. Risk Assessment and Risk Management in Regulatory Decision-Making. Washington, D.C.: RiskWorld.

Vajjhala, S.P. and Fischbeck, P.S. 2006. Quantifying Siting Difficulty: A Case Study of U.S. Transmission Line Siting. *Energy Policy*, 35(1): 650-671.

Wolsink, M. 2000. "Wind power and the NIMBY-myth: institutional capacity and the limited significance of public support." *Renewable Energy*, 21(1): 49-64.

WASTEWATER UTILITY

Alabama
Montgomery Water Works & Sanitary Sewer Board

Alaska
Anchorage Water & Wastewater Utility

Arizona
Avondale, City of
Glendale, City of,
Mesa, City of
Peoria, City of
Phoenix Water Services Dept.
Pima County Wastewater Reclamation Department
Tempe, City of

Arkansas
Little Rock Wastewater

California
Central Contra Costa Sanitary District
Corona, City of
Crestline Sanitation District
Delta Diablo Sanitation District
Dublin San Ramon Services District
East Bay Dischargers Authority
East Bay Municipal Utility District
Fairfield-Suisun Sewer District
Fresno Department of Public Utilities
Inland Empire Utilities Agency
Irvine Ranch Water District
Las Gallinas Valley Sanitary District
Las Virgenes Municipal Water District
Livermore, City of
Los Angeles, City of
Napa Sanitation District
Novato Sanitary District
Orange County Sanitation District
Palo Alto, City of
Riverside, City of
Sacramento Regional County Sanitation District
San Diego, City of
San Francisco Public Utilities, City & County of
San Jose, City of
Sanitation Districts of Los Angeles County
Santa Barbara, City of
Santa Cruz, City of
Santa Rosa, City of
South Bayside System Authority
South Coast Water District
South Orange County Wastewater Authority
Stege Sanitary District
Sunnyvale, City of
Union Sanitary District
West Valley Sanitation District

Colorado
Aurora, City of
Boulder, City of
Greeley, City of
Littleton/Englewood Wastewater Treatment Plant
Metro Wastewater Reclamation District, Denver
Platte Canyon Water & Sanitation District

Connecticut
Greater New Haven WPCA
Stamford, City of

District of Columbia
District of Columbia Water & Sewer Authority

Florida
Broward County
Fort Lauderdale, City of
Jacksonville Electric Authority (JEA)
Loxahatchee River District
Miami-Dade County
Orange County Utilities Department
Pinellas County Utilities
Reedy Creek Improvement District
St. Petersburg, City of
Tallahassee, City of
Toho Water Authority
West Palm Beach, City of

Georgia
Atlanta Department of Watershed Management
Augusta, City of
Clayton County Water Authority
Cobb County Water System
Columbus Water Works
Fulton County
Gwinnett County Department of Public Utilities
Savannah, City of

Hawaii
Honolulu, City & County of

Idaho
Boise, City of

Illinois
Greater Peoria Sanitary District
Kankakee River Metropolitan Agency
Metropolitan Water Reclamation District of Greater Chicago
Wheaton Sanitary District

Indiana
Jeffersonville, City of

Iowa
Ames, City of
Cedar Rapids Water Pollution Control Facilities
Des Moines, City of
Iowa City

Kansas
Johnson County Wastewater
Unified Government of Wyandotte County/ Kansas City, City of

Kentucky
Louisville & Jefferson County Metropolitan Sewer District
Sanitation District No. 1

Louisiana
Sewerage & Water Board of New Orleans

Maine
Bangor, City of
Portland Water District

Maryland
Anne Arundel County Bureau of Utility Operations
Howard County Bureau of Utilities
Washington Suburban Sanitary Commission

Massachusetts
Boston Water & Sewer Commission
Massachusetts Water Resources Authority (MWRA)
Upper Blackstone Water Pollution Abatement District

Michigan
Ann Arbor, City of
Detroit, City of
Holland Board of Public Works
Saginaw, City of
Wayne County Department of Environment
Wyoming, City of

Minnesota
Rochester, City of
Western Lake Superior Sanitary District

Missouri
Independence, City of
Kansas City Missouri Water Services Department
Little Blue Valley Sewer District
Metropolitan St. Louis Sewer District

Nebraska
Lincoln Wastewater & Solid Waste System

Nevada
Henderson, City of

New Jersey
Bergen County Utilities Authority
Ocean County Utilities Authority

New York
New York City Department of Environmental Protection

North Carolina
Charlotte/Mecklenburg Utilities
Durham, City of
Metropolitan Sewerage District of Buncombe County
Orange Water & Sewer Authority

Ohio
Akron, City of
Avon Lake Municipal Utilities
Butler County Water & Sewer
Columbus, City of
Metropolitan Sewer District of Greater Cincinnati
Montgomery County Water Services
Northeast Ohio Regional Sewer District
Summit, County of

Oklahoma
Oklahoma City Water & Wastewater Utility Department
Tulsa, City of

Oregon
Albany, City of
Clean Water Services
Eugene, City of
Gresham, City of
Lake Oswego, City of
Oak Lodge Sanitary District
Portland, City of
Water Environment Services

Pennsylvania
Philadelphia, City of
University Area Joint Authority

South Carolina
Beaufort-Jasper Water & Sewer Authority
Charleston Water System
Mount Pleasant Waterworks & Sewer Commission
Spartanburg Water

Tennessee
Cleveland Utilities
Knoxville Utilities Board
Murfreesboro Water & Sewer Department
Nashville Metro Water Services

Texas
Austin, City of
Dallas Water Utilities
Denton, City of
El Paso Water Utilities
Fort Worth, City of
Houston, City of
San Antonio Water System
Trinity River Authority

Utah
Salt Lake City Department of Public Utilities

Virginia
Alexandria Sanitation Authority
Fairfax, County of
Hampton Roads Sanitation District
Hanover, County of
Henrico, County of
Hopewell Regional Wastewater Treatment Facility
Loudoun Water
Lynchburg Regional Wastewater Treatment Plant
Prince William County Service Authority
Richmond, City of
Rivanna Water & Sewer Authority

Washington
Everett, City of
King County Department of Natural Resources
Seattle Public Utilities
Sunnyside, Port of
Yakima, City of

Wisconsin
Green Bay Metro Sewerage District
Kenosha Water Utility
Madison Metropolitan Sewerage District
Milwaukee Metropolitan Sewerage District
Racine Water & Wastewater Utility
Sheboygan, City of
Wausau Water Works

Water Services Association of Australia
ACTEW Corporation
Barwon Water
Central Highlands Water
City West Water
Coliban Water Corporation
Cradle Mountain Water
Gippsland Water
Gladstone Area Water Board
Gosford City Council
Hunter Water Corporation
Melbourne Water
Onstream
Power & Water Corporation
Queensland Urban Utilities
South Australia Water Corporation
Sydney Catchment Authority
Sydney Water
Unity Water
Wannon Regional Water Corporation
Water Corporation
Water Distribution Brisbane City Council
Western Water
Yarra Valley Water

Canada
Edmonton Waste Management Centre of Excellence
Lethbridge, City of
Regina, City of, Saskatchewan
Toronto, City of, Ontario
Winnipeg, City of, Manitoba

New Zealand
Watercare Services Limited

STORMWATER UTILITY

California
Fresno Metropolitan Flood Control District
Los Angeles, City of, Department of Public Works
Monterey, City of
San Diego County Department of Public Works
San Francisco, City & County of
Santa Rosa, City of
Sunnyvale, City of

Colorado
Aurora, City of
Boulder, City of

Florida
Orlando, City of

Georgia
Griffin, City of

Iowa
Cedar Rapids Water Pollution Control Facilities
Des Moines, City of

Kansas
Lenexa, City of
Overland Park, City of

Kentucky
Louisville & Jefferson County Metropolitan Sewer District

Maine
Portland Water District

North Carolina
Charlotte, City of, Stormwater Services

Pennsylvania
Philadelphia, City of

Tennessee
Chattanooga Stormwater Management

Texas
Harris County Flood Control District

Washington
Bellevue Utilities Department
Seattle Public Utilities

STATE

Connecticut Department of Environmental Protection
Kansas Department of Health & Environment
New England Interstate Water Pollution Control Commission (NEIWPCC)
Ohio River Valley Sanitation Commission
Urban Drainage & Flood Control District, CO

CORPORATE

Advanced Data Mining International
AECOM
Alan Plummer Associates, Inc.
American Cleaning Institute
Aqua-Aerobic Systems Inc.
Aquateam–Norwegian Water Technology Centre A/S
Associated Engineering
Atkins
Bernardin Lochmueller & Associates
Black & Veatch
Brown & Caldwell
Burgess & Niple, Ltd.
Burns & McDonnell
CABE Associates Inc.
Camp Dresser & McKee Inc.
Carollo Engineers Inc.
CET Engineering Services
CH2M HILL
CONTECH Stormwater Solutions
CRA Infrastructure & Engineering
CSIRO (Commonwealth Scientific and Industrial Research Organisation)
D&B/Guarino Engineers, LLC
Damon S. Williams Associates, LLC
Ecovation
EMA Inc.
Environ International Corporation
Environmental Operating Solutions, Inc.
Fay, Spofford, & Thorndike Inc.
Freese & Nichols, Inc.
ftn Associates Inc.
Gannett Fleming Inc.
Garden & Associates, Ltd.
Geosyntec Consultants
GHD Inc.
Global Water Associates
Greeley and Hansen LLC
Hazen & Sawyer, P.C.
HDR I HydroQual
HNTB Corporation
Holmes and McGrath Inc.
Hydromantis Inc.
Infilco Degremont Inc.
Jacobs Engineering Group Inc.
Jason Consultants LLC Inc.
KCI Technologies Inc.
Kelly & Weaver, P.C.
Kennedy/Jenks Consultants
Larry Walker Associates
LimnoTech Inc.
Malcolm Pirnie, The Water Division of ARCADIS
MaxWest Environmental Systems, Inc.
McKim & Creed
MWH
Neotex Ltda
NTL Alaska, Inc.
O'Brien & Gere Engineers Inc.
Odor & Corrosion Technology Consultants Inc.
Parametrix Inc.
Parsons
Praxair, Inc.
RMC Water & Environment
Ross & Associates Ltd.
Sanipor, Ltd.
Siemens Water Technologies
Southeast Environmental Engineering, LLC
Stone Environmental Inc.
Stratus Consulting Inc.
Synagro Technologies Inc.
Tata & Howard, Inc.
Tetra Tech Inc.
The Cadmus Group
The Low Impact Development Center Inc.
Trojan Technologies Inc.
Trussell Technologies, Inc.
URS Corporation
Westin Engineering Inc.
Wright Water Engineers
Zoeller Pump Company

INDUSTRY

American Electric Power
American Water
Anglian Water Services, Ltd.
Bill & Melinda Gates Foundation
Chevron Energy Technology
Dow Chemical Company
DuPont Company
Eastman Chemical Company
Eli Lilly & Company
InsinkErator
Johnson & Johnson
Merck & Company Inc.
Procter & Gamble Company
Suez Environnment
United Utilities North West (UUNW)
United Water Services LLC
Veolia Water North America

List as of 4/27/11

WERF Board of Directors

WERF Research Council

WERF Product Order Form

As a benefit of joining the Water Environment Research Foundation, subscribers are entitled to receive one complimentary copy of all final reports and other products. Additional copies are available at cost (usually $10). To order your complimentary copy of a report, please write "free" in the unit price column. WERF keeps track of all orders. If the charge differs from what is shown here, we will call to confirm the total before processing.

Name ____________________ Title ____________________

Organization ____________________

Address ____________________

City __________ State __________ Zip Code __________ Country __________

Phone __________ Fax __________ Email __________

Stock #	Product	Quantity	Unit Price	Total
			Postage & Handling	
			VA Residents Add 5% Sales Tax	
			Canadian Residents Add 7% GST	
			TOTAL	

Method of Payment:

(All orders must be prepaid.)

- ❑ Check or Money Order Enclosed
- ❑ Visa ❑ Mastercard ❑ American Express

Account No. ____________________ Exp. Date __________

Signature ____________________

Shipping & Handling:

Amount of Order	United States	Canada & Mexico	All Others
Up to but not more than:	Add:	Add:	Add:
$20.00	$7.50*	$9.50	50% of amount
30.00	8.00	9.50	40% of amount
40.00	8.50	9.50	
50.00	9.00	18.00	
60.00	10.00	18.00	
80.00	11.00	18.00	
100.00	13.00	24.00	
150.00	15.00	35.00	
200.00	18.00	40.00	
More than $200.00	Add 20% of order	Add 20% of order	

* minimum amount for all orders

Make checks payable to the Water Environment Research Foundation.

To Order (Subscribers Only):

Log on to www. werf.org and click on "Publications."

Phone: 571-384-2100
Fax : 703-299-0742

WERF
Attn: Subscriber Services
635 Slaters Lane
Alexandria, VA 22314-1177

To Order (Non-Subscribers):

Non-subscribers may order WERF publications either through WERF or IWAP (www.iwapublishing.com). Visit WERF's website at www.werf.org for details.